The Perfect War

Book 1

Apollo Diaz

Dedication

Dedicated to angel. My rain of fire.

About the Author

Apollo Diaz, also known by his given name Gianni Diaz, is an accomplished e-commerce professional, U.S. Marine, and trained psychologist. His life has been full of discipline, learning, and service, and now he brings all that experience into his writing. *The Perfect War* is his first novel and the beginning of a powerful trilogy. With a passion for storytelling and a heart for adventure, Diaz writes to inspire, entertain, and remind readers that even in chaos, there's always hope.

Preface

The Rain of Fire

was an unthinkable disaster that shattered 17-year-old Will Light's world.

Meteors and ash fell from the sky, blotting out the sun and reducing everything in their path to ruin. In the chaos, Will is thrust into a battle between survival and total annihilation as he searches for his sick younger brother in a dying world.

To stay alive, he aligns with a private military group known as FIRA, led by the hardened Commander Dominic. Together, they confront a mysterious and deadly enemy—one barely understood but terrifyingly real.

As danger closes in, both Will and Dominic are willing to risk everything to reach their goals. But for Will, each step forward comes with impossible choices—ones that could shape the future of every life they encounter.

August 18th, 2035.

8:42 AM

Somewhere in Florida.

"Three days?"

"The public needs to know."

"Are you kidding?! The company will be ruined!"

"Doctor Stone. Understand that the whole *world* could be ruined."

Stone looked at the screens in front of him as waves of worry crashed over him. Never would he have thought that he would have been the cause of such a large-scale extinction-level disaster. If the world knew what he had done, they would have his head on a pike.

"Call your family, Martinez, and tell the rest of the ECTO specials to send their final messages," Stone said sternly. "Omen's design of the room will survive an impact if it does happen, but not one word about what's happening goes out of this program. Understand?"

"Yes sir," she sighed in a defeated tone as she stalked out.

"You there! What's your name?!" Stone barked, pointing at a skinny man in his late 20s typing away at a red chart beaming on his face, illuminating his drooping eyes and nervous look.

"Walker... sir?" he said, puzzled, turning around.

"Call my brother! And keep from mentioning anything about this situation!"

"Yes sir," Walker answered hastily while quickly grabbing the phone on the wall next to him. After a quick search in the system, he fumbled with the dial numbers, and a few clicks later, the ring sounded before a heavy-toned man answered.

"Who is this?"

The aggression seemed accidental, but still, it sent chills down Walker's spine.

"Mr. Stone... your brother would like to speak to you."

"Who?"

"Your brother, sir."

"Then put him on the phone!" The man yelled, annoyed.

The doctor snatched the phone away from Walker, "Sorry, Dom! My coworker here is multitasking," he said. His demeanor shifted to easy and ongoing, a sudden change from the once tense and stern man that had stood before.

"Doctor. There seems to be trouble?"

Stone looked at the chart again on the big screens above and below him, furrowing his brows, hoping it was just a fluke— if anything, a hallucination as it continued beaming red.

"A little, but it's nothing that should concern you. I just need you to step away from home. Occupy yourself at work or something if you can, please. Just for the week."

"No problem, doc. I was planning to do some bombing drills with the boys anyway." He replied, *"Should I ask why?"*

"Nothing worth bothering you for, brother. I have some very… abstract people and projects coming home that I can't have any distractions for," Stone said, his cheerful tone faltering slowly.

However, the man was convinced.

"Alright, doc. Just call if you need me to kick some informant's ass." The phone choked out before the phone clicked, and the line was cut.

"Always a busy man." Stone thought to himself, tossing the phone back to Walker.

As the clock struck nine, Martinez, a slender woman, strutted inside the room. "All our specials are notified, sir. Although, Omen has not sent a response back to our call."

"Thank you, Doctor Martinez," Stone said without looking up.

"What now, sir?"

Stone again hoped that the doomsday counter above was a fault. A sensor misinterpretation. But as the minutes passed slowly, one after the other, all he could do was answer truthfully.

"We wait. Hope and pray."

August 21st, 2035.

5:30 AM

Abilene, Texas.

"Can days go by any slower?" I wondered to myself as I woke from a long, dreamless night, which had been happening a lot more recently. I didn't know why. But other than that, nothing changed.

Anyways. My name is Will, or technically William Lucifer Light. Yeah, I know. Devil's name for my middle name, but I never really cared. I've never really been big on religious Christian Gods and all that — mostly a science kinda guy. Evolution and boring stuff like that. Just a highschooler with a common antisocial personality. Is that even common, though? Not sure.

When I woke up, I realized a few things.

- First, my hair was a mess.
- Second, my room was a mess.
- Third, my life's a mess.

I try not to seem ungrateful since I was born in America, the land of the free, and that I have great parents. Excuse me… *Parent.* My father died a hero in the army three years ago, yes. Before you ask, sometimes it still upsets me, hell it would upset anyone if they lost their dad, even if it was five years ago or more. He was a good man, strong, brave, and a good-looking one with a nice sense of humor to match. But what I really missed was his presence. I just wanted a father again, one that would teach me all the "Growing Up" things, such as how to drive, get a job, or use a gun. Someone who could shed light even in the darkest of times and magically fix almost anything. One like my father, Thomas Light. It's sorrowful, but eventually, we all need to move on at some point, I suppose. My mother, Arabella Light, is still here. She is a strong resilient woman even after everything she's been through.

5:32

Damn. A whole hour until I actually have to wake up.

As I got up, I started my usual morning routine.

Get out of bed, trip over my dumbbell and curse, get up again, grab school clothes, go to the bathroom to brush my teeth, take a shower, and comb my hair. Do I have to go on?

As I dress, I wear the least noticeable thing in a Texas military high school that a student can manage to get away with. Same thing as a regular Texas high school. Only… military structured.

A black skin-tight shirt and blue jeans with black combat boots were my usual. Of course, I can't forget my prized possession, a brown leather cadet trench coat with four pockets on the outside of the leather. It came in handy a lot more than you'd think it would. It was a hand-me-down from my father himself. It's nice to have something of his with me at all times. The rest of the stuff he left behind when he died was sold or given away due to our living situation. All we had left really were some clothes, his 1911 black star, and the armory gear he used while serving. We were given the gear after his sacrifice to the country.

It should be on display, in my opinion.

I headed out of the bathroom and made it downstairs. It never fails to surprise me when I see my mother awake before me. Even this early.

"Good morning, William," she said in a tone as soft and easy as it was years before.

"Mornin' mom."

"Did you sleep?"

"No," I answered sarcastically, as she smiled and shook her head.

It is always a warm feeling when you get your mother to smile.

"Don't fall asleep in class," she joked.

I smiled and looked at her. you couldn't even tell her age. At the ripe age of 42 my mother looks like she is in her prime effulgence youth, with dirty blond hair with little to no silver spots. She was a little tall — that is, if you count 5'10 as a little tall. I'm 6ft, and she still can look intimidating for her size. She usually dresses professionally when she is working at the storyboard, but now she wore a navy blue dress with yellow blossoms spiraling on the skirt — one she got at her wedding.

"Are you going to eat here or at school?" she asked, snapping me out of my thought.

"I'll eat at school," I answered warmly.

I would love to eat at home, but again, money was tight. We didn't need extra money going to food that school provided already.

She leaned on the table. I could hear coffee being brewed by the senile coffee maker. It could retire any day now.

"How did you sleep then?" she asked, trying to make the usual quick talk.

"Good."

"English William."

"Divine and quickly." I said, trying to sound pleasant.

"Good. Now when you get home, I may not be here. I'll be visiting your brother today so please prepare yourself something after school."

Ah.

My Brother.

Where do I start with him? It's a touchy subject as if it would be to anyone if they never got to see their younger sibling for years upon end. I saw him leave when I was 13. He was three at the time. Now, he was around seven to eight. The reason being that he had a rare disease. A very rare one, in fact.

At least, that's what I was told.

So, in my mother's words, he is at the hospital being treated, and visits are only allowed by the parents. He was not even allowed to talk via phone to me. When I tell you I would *beg* to visit Jack, just once, only to see if he was okay, take my word for it. Now, I just assume he passed away from it, and my mom, Alice, doesn't know how to tell me. I sometimes have the intrusive thought to go inside his room, to see what it's like after all this time. But it's set to key lock by Alice. So that was her way of saying, "Keep out!" I assumed she took a night shift at the press. So, I took the bait, just so she didn't worry more than she had to.

"Okay, Mom."

Ding.

That was the sound the decrypted coffee maker made when it was finished. She got up and took out the hot clear mug, and she poured the coffee into her flask.

"Oh! Could you take the trash to the curb when you leave? It just crossed my mind."

"Yes, of course," I said as she headed out the door.

Great. Alone. Again.

Awkwardly, I moved to the kitchen trash can. Taking out the half-full half-empty bag. On the doorway out was my bookbag, which really was just a computer bag. Of course, I didn't have a computer. Hell, I didn't even have a phone.

You know… Money.

I grabbed the bag, hoisted it on my shoulder, and headed outside. While I put the trash bag inside the trash can, I saw a young couple walk by. Both looked no older than 16.

Young love. Ew.

Sure, it's cute, but in my opinion, it's almost useless. I'm 17, and I know that this could end catastrophically. Like why find love at a young age if it's known to be bound to a breakup? I thought this as I hauled the trash bin behind me to the curb. Yeah, in case it wasn't obvious, yeah, no. I don't date, and I don't think I ever will. Maybe if it was ten years ago, then maybe, but some relationships at school that I see last no longer than a month. The longest I even heard of was two months solid. Love's complicated now. Everything about you is digital, and anyone can get to know you without even talking to you. Or maybe it's complicated for me. Yeah, sure, I thought of it but never took enough interest. Plus, there is only one keyword in my life: *Antisocial*, Remember?

I placed the trash bin by the curb and returned to lock the door. Only to realize:

"Shit. I left them inside," I exclaimed out loud.

I rushed inside and headed back upstairs to look for my wallet. Why my wallet? My keys were attached to it. Why? Because management. The brown leather wallet was found on the floor of the bathroom, and my three keys were attached to the zipper clip. This was why I attached them to my wallet. So, what if we were money-tight? I still needed my wallet. Not much was in it, though. There was an old digital Fair sticker, an army recruiter business card, a simple, thin digital pocket watch, and two slightly ripped 20$ Bills. But I never took my wallet for the time and money only. I took it for the last item in it. A photo taken in 2030, on Christmas day, with me, my brother, and my parents, I was 12 then. There I was in the picture, smiling without a care in the world. I closed my wallet and walked back downstairs, out the door, locking it behind me and shoving the wallet in my pocket. The last time I read on that clock was 6:18.

Perfect. I had enough time to mess around in the forest.

August 21st, 2035.

6:22 AM

Abilene, Texas.

I headed into the wall of trees and bushes that the forest began in. Who knew how long these tall redwood trees went? Acres could stretch for miles, and God knows what could be in that underbrush. As I traversed the rugged landscape, I kept an eye out for wolves.

Yeah. Wolves.

Fall was beginning, and if I hadn't spent my time looking those following few days, I might not have been able to see one until spring due to their hunting patterns changing to survive the frigid cold. They did seem more active during the winter. Just less noticeable, like me.

I followed the trees quietly; wolf or not, if I made too much noise, I would startle whatever lurked around. I didn't need anything stalking me too.

Oh yeah,

Military school Not by choice. It was the only free school around. Plus, it would provide more for me when I grew up and joined the army. Just like my father.

I went to Churchill Military Academy, named after the famous Jack Churchill. He was a hero for sure, but to me, he was more of a badass Rambo with a gun. When you're the best soldier in a platoon, they don't call you the best. They call you a Mad Jack.

It wasn't that bad. Sure, the drills and PT were annoying, but they would at least teach you right. Or rather, I paid attention the best I could because every little bit of information was essential, training or real scenarios.

As I continued to walk, I noticed more and more droppings, not wolves but deer. Which was also good because if you saw deer poop., then there was deer. And where there were deer, there were wolves. I continued checking my watch along the way. It was still 6:40.

I had to give up. School would start in about an hour, and I was buried deep in the woods. I turned around, defeated.

Snap!

Then I heard a yelp as if a dog was getting kicked in the gut.

Forgetting the time, I rushed to the whimpering sound that followed. It *had* to be a wolf, but how was it hurt? And could it not run off? If so, why? These thoughts flew by as I rushed to the growing crying of the poor animal until I saw it.

There it was, a gray-furred wolf — a giant one, in fact. This wolf looked bigger than me! I studied the wolf cautiously. Its feet stretched out, and its eyes narrowed while its muzzle pointed down, signaling that it was ready to attack.ready to attack me.

His left back foot, although, was stuck in a dull bear trap. No spikes on it, lucky enough, so it was not meant to kill. But to trap. While I inspected the trap from a distance, the wolf growled at me.

So, I was faced with a choice: to either help the wolf and risk getting mauled or leave it for the trapper.

"Hey buddy… easy…" I said softly, sticking my hands out to show and signify that I was not a threat. The only response I got was a louder growl, with it shining its canine teeth.

I grabbed a stick and tried prying it open with it. The wolf stopped showing its teeth at me, but it continued its loud, dangerous growl. I gradually came closer, trying to jam the stick in between the jaws of the trap to keep it from closing more, only to have the stick snap. Fuck!

I then stuck my hand out, ready to release the trap myself.

Before you call me stupid, just know that I was running late for a *Military* School, which was *never* good.

Never.

I reached closer and closer, hoping not to get bitten or worse. Its intimidating growl grew. I had to be quick.

So, I quickly grabbed it and sent the wolf free. While I did, I heard the beast snap.

August 21st, 2035.

6:59 AM

Abilene, Texas

The wolf didn't snap at me, (Thank god.) it snapped away when I freed it. At the moment, I didn't know where it went because I was busy covering my neck. Literally, I was holding my hands over my neck so if the wolf did charge at me, it would attack something less vulnerable. I would much rather lose my hands over my neck any day. It was a tactic I learned from my dad.

I poked my head up and saw the hound standing there, menacingly. It was about seven to eight feet away from me, and it stood… staring.

I would usually get stares, but none like this; I picked myself up and cleaned off my coat the best I could. Sgt Parlor would bury me alive if he saw how dirty I was.

Sgt Parlor. Oh shit.

I quickly finished brushing myself off, and I looked at the majestic animal that stared back.

"Gotta go, bye."

Then I turned and ran off. The wolf did the same and headed to wherever wolves frolic too at that time. As I did, I fumbled, pulling out my watch. It showed 7:02.

I had to be there by *7:30*, and it was a 50-minute walk! With no choice again, I had to run. I rushed through the forest. Tree branches, bushes, sticks, and, um… whatever else grew in the way had to be shoved over by me bulldozing through.

I finally saw the sidewalk and busted out of the underbrush like a crazed animal.

"Glad no one saw that…" I mumbled to myself after a quick perimeter check, still dashing down the path.

I continued running until I was out of breath, huffing and puffing as if I had smoked three packs of cigarettes. I checked my watch again. It was already 7:15.

There was no way I was going to make it. I still had a 25-minute path, and I was exhausted.

Even if that was the case, I still could not give in.

I kicked to full sprint; if I kept this speed, I'd be able to make it before the bell. And my death sentence. I made good time, too, but the sprint was killing me. I ran and ran until I reached the end of my street. When I did, I could spot the school clock to the left of my road at a distance. Still far off, yet I could make out the time. 7:26

Shit.

Shit. Shit. Shit.

I ran with all the hidden energy I had left, feeling like dead weight already. The adrenaline was killing me — my feet, screaming for a break.

"Never give in."

Those annoying words hit me again. My father said it once, and now it popped up in the most annoying of times.

"Never give…"

"Yeah, Yeah. Quiet now, please!" I almost tried yelling in my mind. Annoyed

But I continued sprinting, getting closer and closer by the minute until finally…

"Never give in."

"Okay! Happy?!" I yelled out loud in front of the whole academy of students rushing to class. I got a few odd looks, but I didn't care. It was not over. It was 7:29, and my class was starting. I rushed to make it to a classroom that was on the other side of the school.

"Hey!"

"Sorry…"

"Watch it! Private!"

"My bad…"

"Don't be late, wuss!"

"Fuck off!" I yelled running past crowds, bumping into a few people on accident. Some people really didn't like me after I did.

Anyway, I reached the classroom as the bell rang and walked inside.

"Private Light!"

I looked over to see him behind the desk, standing with his arms crossed. Great.

"Good morning, sir!" I fell at attention quickly.

"As you were."

I spread my legs out and put my hands behind my back. Everyone sat in their desks inside were now looking. And Sgt Parlor stood there, as menacing as the wolf. Except rather than a gray-furred mutt, it was a 6'7 dark-toned man. Slightly balding, yet still in rough and tough shape. While all the students stared, I kept my eyes forward as the Sgt kept his eyes on me like I was a nice juicy T-bone that he couldn't wait to poke and play with before devouring.

"Where were you Private?!" He demanded. Very, very loudly.

"I was working out until I lost track of time, sir!" I answered with a choke.

"Read me the cadet creed Private!"

"Yes, sir!" I answered back.

"Cadet Creed

"I am an army military cadet; I will always conduct myself to my family, country, school, and corps of cadets. I am loyal and patriotic; I am the future of the United States of America. I do not lie, cheat, or steal, and I will always be actions and deeds. I will always practice good citizenship and patriotism. I work hard to improve my mind and strengthen my body. I'll seek the mantle of leadership and stand prepared to uphold the Constitution and the American way of life. May God give me the strength to go by this creed."

If you skipped over that whole speech, I don't blame you.

"Mhm, and where did you say you were, Private?"

"I was working out at home… sir," I said, shaking.

"Read me the cadet creed again, Private!"

I heard some low chuckles from the students.

"Cadet creed…" I sighed.

"Sound up!"

"Cadet creed!" I said, louder.

I repeated the cadet creed until he held his hand up to stop me.

"Say what you just said again."

"I do not lie, cheat, or steal and will always be accountable for my actions and deeds," I responded.

"Uh, Huh, okay. I'm going to ask you again," he said with his eyes narrowing,

"Where were you, Private?" he asked.

Though it sounded a lot more like a demand, damn, he's good. Maybe I had a bird in my hair, like in those old movies.

"I was in the forest freeing a trapped wolf," I responded softly.

"Sound up, Private! I don't think the class heard you," he said back, almost softly, and… dangerously.

"I was in the forest treating a trapped wolf, sir!"

Of course, the class laughed. Not for me, or with me, but at me. It felt like I was in front of a firing squad.

Or, for those of you who don't know what that is, I felt really, really judged.

"Read me the cadet creed again, Private!" He barked.

"I will sir… But…"

"Private!" He yelled.

I continued to look forward in silence, hoping not to have dug a deeper hole.

"Whatever story you give, I won't care for! Lies or not, you're still late! Understand?"

"Yes, sir…"

"Sound up!"

"Yes, sir!"

"Go to the office and get a late tab. Fall out."

I stepped out of position and quickly made an exit, restraining myself from slamming the door. Then I heard a low murmur from inside.

"Boy can't lie himself out of…"

I walked away as I heard the class burst out laughing again.

I wondered what the joke was. I would love to hear it.

August 21st, 2035.

7:39 AM

Abilene, Texas

I kinda understood why he didn't believe me. It was a crazy excuse, too. Like Me? Save wolf? Not get hurt? A few seconds late? Nah.

While I was walking down the hall, I noticed out of the window it was a lot dimmer outside. Possibly, a storm was forming. That was good. This meant no PT today, and Sgt Parlor wouldn't get a chance to go so hard on me for the rest of it.

I reached the front office door and walked in, taking account of the parent who sat in the waiting chairs and the desk lady clacking at her loud, loud, *loud* computer.

"Mornin' honey," she said in a tone as old as wine and sweet as pie. Always so friendly. I didn't know her well, but she was always on one of my "Teachers I like" lists. Not that I had many of them. "Need a late pass?"

"Will Light, Yes ma'am," I answered.

"Okay, Hun, one minute," she said, clicking her mouse a couple of times and typing seemingly randomly. I only noticed a few key features from her when I looked. Short, thin, dark, clean skin, she looked like a woman who would be singing blues on stage instead of a lady taking names and giving passes to high schoolers.

Tic Tic Tic choked the printer as it printed my late tab.

"When it's done, can you please turn it off?" she asked.

"Of course. Thank you, ma'am."

A long, 20 seconds later it finally finished printing. I grabbed it, pressed the power button at the top, and walked back towards class, ready to deal with my embarrassing stand-up. Suddenly, though, the intercom came on.

"At this time, we are having a tornado drill. There have been some unexpected weather readings, but please follow the instructed drill protocols. I repeat. We are in a tornado drill."

Then, the click of the microphone was heard.

Okay, I had two sides on this. The "Ugh" and the "Yay." Ugh, because now I gotta duck and hide somewhere, or else I would get chewed out by the Sgts. But yay, because, no more classes.

I looked around and saw the janitor's closet in between lockers. I opened the door handle (thanking the universe it was unlocked) and went inside, flicking the single lightbar on and closing the door. I did not want to crouch on the floor with my hands over my head like I was a turtle. Is it wrong to find that annoying? I mean, sure, it's for "Safety," but like, what's turning myself into a *beetle* gonna do against 300 mph speeds of straight *air and wind?*

I just leaned against the shelf and waited. It was not a big room at all. I spread my arms out and I touched both walls at the same time *with ease.* So, instead of getting comfortable, I just tried to name all the different scents I smelled.

Pine-Sol, degreaser, glass cleaner, alcohol, wood cleaner, and dawn soap were the less obvious ones. However, the room reeked with the smell of bleach the most. And I hated it. It made me feel lightheaded a little.

The kids in the hallway started walking outside. How did I know? I could feel it. The vibrations of their stomps going down the hall past the closet.

Until it got harder.

And harder…

And louder…

And rougher…

Holy, was there a stampede out there?! Why were they all running? Moreover, why were they running into everything around *While* running?

Then, the whole room started shaking like an earthquake. The cleaners were all on the shelves, tipping, falling, vibrating. The light flickering on and off for who knows what.

I was scared. I tried to open the door…until the door… broke off its hinges. A loud crash followed as the door flung down and hit me on the head, knocking me out cold.

?????? ?? ????

??? ??

Abilene, Texas

I woke up to a blurry vision. Perhaps it was due to the light. The only thing I could feel was the solid iron blanket covering me.

I pushed the door over and was conflicted about what I saw. Very conflicted.

The sky was gray with a very dim shade of… purple.

Purple?

I sat up and looked around. The air smelled horrible. As if the Fourth of July went off for a whole week — the air reeked with gunpowder and smoke. I looked down while sitting and looked at myself, covered in dust, but it was not just me. The surrounding area… the academy, was… ash and rubble. Not one recognizable frame stood of the school. In the middle of it was a crater. And inside the crater was a giant rock. And it was… a smoking one. I got up and checked myself for any injuries. Aside from a few cuts and the feeling of a bruise building on my neck, I was okay.

Finally, a minute later, my mind kicked into full reboot.

"What the hell?! What the hell?! *What the Hell?!*" I screamed.

"Hello?! Is anyone out there?!

No response. Great.

Just wonderful.

I worked my way around the rubble until I reached the boulder. It was as big as a Mack Truck (with no trailer included) and it had holes in it, like a sponge. I stuck my hand out to it, but I avoided touching it. Pure goosebumps avoided that. The dive and the looks of the rock helped me figure out what happened. My school was hit by a meteor.

Let me repeat. A *Meteor*.

Hah.

I had to be dreaming. First, I save a *wolf,* and now a *meteor* hits *My* academy?

I'm dreaming.

I'm hallucinating.

I'm insane.

I'm crazy.

At least, those were the thoughts that kept running in my head the whole day.

"They probably have me in a white box now. I'm just shaking inside, rocking back and forth violently on a solid stone floor," I thought to myself.

Checking myself again, I was in rough shape. My clothes were dusted and had holes in them, and my dad's coat was ripped from my beltline, so it was more of a jacket now. My shoes, although, were fine.

Steel toe baby. Gotta love them.

But with that came the realization that all this was *real*. This wasn't a vision of me inside some new VR gear. The immersion, the pain, from the cuts to the bruising, it was all real. Too real to be a dream of some sort unless I woke up any minute now. This was real.

I had to go home.

I started traversing the forest of rubble and steel until, I guess, my adrenaline wore off, and I felt how much my legs hurt. Probably from the run from like… an hour ago? Possibly a little longer…

I reached the office… *ahem*. That was… where the office *used* to be. I stopped after my legs screamed in pain from that little march.

There was no way I'd make it all the way home. In this condition, I'd faint on the street. My legs felt like they were smashed by a building.

Huh… I wondered *why*.

But not all hope was lost. Under me, I noticed a solid silver box, a key box under a roof tile. These keys were for the Humvees' and trucks that the school used for drills and training and whatever else. I could see the car lot on the side of the academy from where I was standing. I could make out the Humvees' and the Mack's, and only a few were… flipped.

I grabbed the box and opened it. Grabbing a random key that had a red tag on it. Dropping the box, I walked out of the rubble and headed to the lot, dragging my feet on the dirt, which should *not* be dirt. The grass was dead. Hell, everything the eye could see was dead. I didn't even see anything. It was all… *Green*. It's like the meteor killed *everything*.

On the keychain of the key was the red tag that had a plate number on it, along with the row and the designated parking spot, which was "C-13." So, I headed to one of the first rows of cars. Walking to the row, I saw the hummer of C-13. It was in good shape, but I wish I could say the same for the others. Some —like mine, had holes inside them the size of baseballs. Some were bigger, others were flipped, and one carrier truck was crushed by a meteor. My school was hit by *two* meteors! How unlucky can you *get?*

I found my way to the door and started the monster up. Full tank, *thankfully*. Plus, there were an extra two canisters on the back bumper. But the downside was that these had horrible gas mileage. I'd be lucky to reach the end of the state.

I didn't need to worry about that. I was not going to China. I was only going home. I carefully drove out of the lot and into the street. Driving down the street, I noticed the worst possible thing. Well, okay, one of the worst-

It was not just my school — but shops, cars, and even the *streets* were destroyed by meteors. It wasn't just my school! It was the whole city. I realized that after I saw the office building on the side of the street.

"It's the States, so the bank probably has it covered." I thought to myself. I wonder how long this goes on for...

Could this have happened to other towns, cities, or states? Maybe all the ones around? Where was the police? Or even the people? Did they get hit, too? Did the whole country get hit?

"No." I thought as I shook my head, "I can't worry about that; I gotta go home *and go get Alice."*

The ride was steady. Apart from avoiding the rocks and abandoned cars, all was smooth. Two and a half miles was nothing with a car.

I did try turning on the radio to calm myself down, but I should have known it was not gonna work when I drove past a crashed radio tower. Knowing it may be hopeless, I tried talking on the cadet cabinet radio. It was already connected to channel two as I grabbed the microphone and pressed the gray button.

"This is Will Light of the Humvee C-13 Channel requesting aid. Does anyone copy? Over.

Silence.

"Does anyone copy? Over."

Again. *Silence.*

There was no need to dwell on it. I finally made it to my street. Pulling down towards my house, I was me to my core when I saw it. It was standing. Kinda.

?????? ?? ????

??? ??

Abilene, Texas

The roof of the second floor had a giant hole in it. All the windows were busted, too, but aside from that and some small holes, it was intact.

I parked in the driveway, got out, and rushed to the door. I tried to turn it in, but I remembered I locked it. I grabbed my wallet and shoved in the house key.

When I walked inside, I expected more than a few smashed plates and dust-covered furniture when I looked inside. I had to ensure that Mother was okay, so I called out her name.

"Mom?! Hello?!"

The only response back was the house settling. So, I began walking through the halls.

"Mom?" I called again, going into her room.

Gone.

"Shit. Okay, she is not home. No biggie, she still could be okay," I thought as I sat on her bed and recollected my thoughts.

I saved a wolf. I got to school almost on time, a tornado drill happened, I got knocked out, and I woke up to meteors that hit my school, house, and town. A lot went down in a span of a day.

"There is no way…" I said out loud to nobody. "No way."

I lay back and checked my hands, my dusty, dry hands.

This is not a dream. This is not a game. This is real.

I shot up out of the bed. I couldn't let myself get lost in my thoughts. People go crazy when they find themselves getting lost in their thoughts. The government, the army and all the concerned personnel must be all figuring it out. Alice did say she would be out late, so she must be… somewhere safe… maybe. She could be with my brother, who could be alive. Or maybe at the office…

I went into her closet because I knew one thing that was in there. Pulling out a large digital camo bag, I put it on Alice's bed. It was Dad's gear.

Inside were fresh cargo pants, a cargo belt, A heavy Kevlar vest, a camo helmet, and his 1911 black star. The one thing that probably saved my father countless times was probably going to save me.

I changed my pants to the camo ones and put back the Kevlar vest and helmet inside the closet. It was a meteor attack, so I did not need to think I would be getting *shot* anytime soon. I just wished I had a holster for the gun. You know, in case I *did* get shot anytime soon.

I stuffed the gun inside one of the front pockets of the bag. I felt something more than bag. Opening it and pulling everything out on the bed. The items that were laid out were two flares, a compass, a small trauma kit, two ponchos, a canteen flask, and a six to seven-inch combat knife. I unsheathed it from its leather cover and inspected it. It was fully black, stainless steel, with four inches of the blade having a saw-like pattern. Sharp, deadly, and very beautiful. At least I could hold on to this.

I clipped it around my new belt on the back of my hip. I also stuffed everything else back inside that small pocket with the gun and tried exiting the room, leaving the bag on the bed. But, when I walked to the doorway with that, I saw a glint at the top of it. Sliding my hand across it, I felt something small and metallic.

A key lingered in my palm.

A key?

A Key!

I got back and grabbed the bag. Hoisting it on my shoulders, I ran upstairs, past the bathroom, and past my room.

I had to know.

When I opened the door, my mind felt lighter. The burden that clung onto my chest almost vanished when I saw the blue baseball bed with matching blue shelves that I remembered from long ago. Aside from a few drawers wide open and some boxes in the back corner, it felt close to perfect. The boxes probably being our extra storage.

I walked up to the shelves, on the other hand, and looked through them one by one. There were just papers, drawings, toys and clothes without the slight mention of the top middle shelf having mail in it. I pulled out the small stack of envelopes and read through them. The dates were… recent. This was not old mail. Of the five envelopes, three were open. Two from my dad. And one… urgent hospital notice from August 20th. The paper read "UT Southwestern Medical."

Before I read my father's older letters, I opened and unfolded the paper inside the mail that came from the hospital.

Dear Mrs Light,

August 20th, 2035.

    *It is with due regard that your son, Jack Light's condition has worsened during testing. We will be sending him to our new facility where proper equipment is stored. The location mentioned at the bottom of the letter shows a more appropriate and safer environment for your son. We apologize for the inconvenience of all this. We need you to come in as soon as possible. The unnamed infection that has affected Jack is high kept with medication from being contagious, but we ask you to come in as clean and tidy as possible. Contact our telegram at *64 if anything comes to a line where you can't make it. We swear and promise to you that we have and will continue to try our best to help your son and come up with an antidote or cure.*

With regards,

Doctor Fauci

Doctor Fauchi. August 20th at 9:00 AM

Arlington Rd, Dallas TX 75501

Yikes.

That stung.

Of course, I know I didn't get the worst of that jab. Alice must have been twitching reading this. No wonder she left so early and dressed so well.

But… Jack! Holy shit! He is alive! Like actually alive! And I even have a location! I'll say by the zip code, it's in Dallas, and it says Arlington, so about an hour to two drive. No problem! Plus, I needed to find some help, and because Worth is on the path with Dallas, I figured people would be there, I mean it's the biggest city in the state and the states that surround us. Someone's gotta be there.

I take off my bag and shove the three letters inside the water bottle holder. I'll read them later. I carried the bag out of my brother's room and into my room. Packing essentials that included a fitted set of clothes, a sleeping bag, a lighter, and a book called "The Hunger Games." For entertainment. This brought me to what I needed to pack next. Going to the kitchen, I grabbed food. That's the technical term for it because all that was there was just canned fruits, beans, and bread. Ramen and two smaller pots, to make clean water. Oh, and lots of water, too.

It was time to go. I walked out of the house and stepped back into the Humvee. Putting the bag in the passenger side, I pull out the hospital letter again. Looking at the location

"I need a map." I sigh.

Luckily, there is a gas station right by my school. I could possibly find a map there. And while I'm doing that I could grab more supplies, food, and gas too. Then, I could head on to my trip. Alice had to be with my brother. And I'll see them both tonight or tomorrow. If nothing happened to them, that is.

"Stop being so morbid. Go," I told to myself as I pulled out a key from my pocket, only to realize it was my house keys and my wallet. Looking inside, I saw the digital pocket watch, and it read 6:40. I was knocked out for, like, eight hours?! God, anything could have happened. Worried, I started the old beast up and pulled out of the driveway, driving onto the road.

The destroyed, messy, sweet road.

Same Day?????? ?? ????

6:40 PM

Abilene, Texas

I wish I said my goodbyes to the sun. I never thought I would have to until the sky blotted out all traces of its existence, a horrible omen. But enough light still shone through the sky. So regardless, if I could see, I could keep moving.

I drove on the road, passing all the wreckage again, and made it to the school with my Humvee unscathed and headed to the other side of the street in front of the school, where the gas station, unlike everything else around, kinda stands. The glass was all shattered, and the overhead cover was laced with holes. It was bad, but recognizable, at least. I pulled up under the cover and parked next to a pump.

It felt darker, so I checked my pocket watch again.

6:54

I'll have to camp here. The shack stop will at least hold for the night because I won't be able to drive in pitch black. The clouds will cover the moon too, making light a scarce resource.

I opened the door and was about to step out, until I remembered to turn off the hummer, which I did next. Like I said, conserving gas with it was horrible. And i have no idea how to take gas out of a broken pump. So i mark that off the list of things i needed to do here.

Stepping out and smelling the horrible air again. I attempted to convince myself that I will get used to it. Hopefully.

I walked up to the dusty door and stepped inside. It was no better inside than out. All of the shelves were almost empty. Rather, the food was stolen or on the floor. Everything was covered in a mess which was all over. Some boxes of random snacks were open. The air was no better inside than it was out. The seasoned gunpowder scent lingered with the smell of gas and lead.

So I make myself at home. Kicking some open foods away from an area on the floor where I would sleep. Then I walked around, looking for anything useful.

As I looked around, I tried to distract myself. It was no use. My hands were busy, but my mind had way too much free time. This is all just so bizarre. Crazy. All of this tumbled on top of me like a pile of bricks. And now I was lost in a dead-ish world, with no one else around.

Not even the dumb bugs are around. I wasn't swatting away flies or killing cockroaches. They were dead already.

I grabbed nothing but a map and a flashlight. Putting the bag down in the area I cleared, I shoved the bag and flashlight inside with clenched teeth. After that, I sat down and tried to calm down.

I go inside my bag and grab the lighter. up a few small boxes and small random pieces of wood that would burn into a little pile. Yeah, yeah, I know, Smokey the Bear wouldn't approve of me lighting a fire indoors on the floor. But it's not arson if I'm trying to *survive*.

I lit a box and blew on it softly. I remember I found a bag of coal in the corner, so I got up and dragged it over, throwing a few coal rocks inside.

The fire grew brighter while the outside world got darker. I put a can of beans by the fire, and after about 20 minutes, I served myself.

Bon appetit. I groaned.

I ate the mushy black beans and threw the can away. (By away, I mean away from me) I took out the sleeping bag, having the book fall out while I did. So I started to read it. The fire kept a dim light on it. I read the book before. However, that was like when I was *very* young. The key details I remembered were that the main character gets chosen to fight other kids to the death by the senate of that universe, calling it a "Sport" for a reminder of "The Dark Ages."

Hours passed with little movement, and after a little while, I got tired. So, I marked the page with a piece of burnt cardboard and closed the book. I was already at the part where Katniss was on the train with the drunkard. I could only imagine the stress she suppressed from the reaping. I stuffed the book inside my bag and unrolled my sleeping bag, stuffing myself inside.

I didn't realize how tired I was. I passed out before the fire came close to burning out.

?????? ?? ????

9:20 AM

Abilene, Texas

I woke up woozy today. Rather, from the air inside or the air outside, combined with the air inside. So I shakingly woke up and stood on my feet. Feeling the blood in my head instantly disperse and stars form, I felt dizzy.

Looks like it wasn't a dream after all.

I mean, it definitely didn't *feel* like a dream. These bruises were killing me. I glumly picked up my sleeping bag, which now had an annoying hole on the side from the fire melting the plastic. Luckily, though not melting through me. But it left more unneeded smoke, adding to the already horrible air was not a fun smell.

As I packed, part of me wanted to stay. I was battered and tired. I wish I could cry, but I was already trained enough to shut down my emotions that I couldn't turn them back on when I was in bad times. Plus, I think I'm even dirtier than the air, not that it really mattered.

I had to find Mom and Jack. So, fuck rest. I shoved the sleeping bag inside my cargo bag and picked up the backpack, which seemingly got magically heavier in the night. Strapping it to my shoulders, I noticed the dusty monolog clock above the exit door. And behind the dust, I could make out the time somewhere around 9:35.

I wished my farewells to the campout of the century and walked outside. Pushing the dusty door, I did wonder how the door survived, but the windows didn't. So, just to see, before I walked out, I threw a charred coal in the middle of the clear door. It bounced off with ease.

Plastic. Smart.

Wow, that light hit me. With no sight of the sun, it still did cause my eyes to shut, looking directly up. I squinted my eyelids, waiting for my eyes to adjust, walking to the Humvee. When they didn't, I flung my door open and sat inside, slamming the door.

My eyes felt much better inside. In fact, they didn't feel like they adjusted much from the dim light outside. But looking at the windshield made me realize why. Dust covered it, so I figured that light didn't hit me. Dust did.

By mistake I sat in the passenger side. I guess I was always used to it. My dad was usually the driver, so whenever we would actually stop for gas, it almost never failed for us to go inside and get ourselves three musketeers bars. He was as much of a chocolate lover as I was. Good times.

Those days fell a long time ago. I just wish I could pick them back up.

The key was already in the ignition so another good news-bad news case. The good news was that no one was around to steal my car. The bad news was… you get it.

It rumbled five to six times before it finally roared to life. I turned on the windshield wipers (which didn't really do much) and shifted to drive. I was glad it was a manual. Otherwise, I would be fucked figuring out an automatic. I pulled out of the gas station and into the road, stopping quickly to take off my bag and get the map.

Currently, I'm in Austin. Dallas was a good four hours away. Keeping a good pace with no red lights (gotta watch out for those.) I could be there by *1:00.*

Perfect. Let's see if we can get there before that.

?????? ?? ????

10:19 AM

Capital Interstate Road, Texas

An hour down the road I start to get a little excited, jumpy even. I'll finally be able to see my brother, that is, if he is… alive after all this.

Shaking my head at the thought, I tried to focus more on the road. The destruction down the route was no better than it was around home. Empty streets with the occasional abandoned car. Or a fallen meteor. Even passing a few bodies. Of course, I couldn't stop for them. They were dead. But I hated looking at them. It was hard to drive by them when I first saw them. And still no *living* soul in sight anywhere.

Some of you would think being alone is heavenly, which at certain times is true. Like during schoolwork or whatever, but until you completely lose the ability to have social communication, you won't be able to realize that a desolate life like this is hellish to the mind. It didn't take long for me to feel lonely. I just wanted to see someone, an old classmate, a neighbor. Hell, even Sgt Parlor was better than nobody, even if he was yelling at me for every little mistake I made if I could hear it now one more time.

"Private! Who taught you how to drive!?"

"Private! You almost hit that meteor! Are you Blind?!"

"Private! You said the radio commands wrong! Drop and give me 50!"

As I relay his annoying commands, he would be shouting at me if he was here. I then spotted a sign on the road ahead leading to a turnoff.

"Turnpike to worth - 80 Miles"

The thing is… there were holes in it. And it had bent crooked a little due to an old Honda that seemingly crashed into it. Holes inside the car, too, did not seem like they were meteor holes.

I slowed down and parked on the roadside, digging into my bag and pulling out my gun. The area just got a lot more dangerous.

I shoved the gun in my beltline behind me and stepped out. And walked up to the car, where I … a person

God, it's a person! And he looked unconscious, too! His head below his hands on the steering wheel.

Maybe he got knocked out from the crash. I ran up to his door side and opened it, pulling him out.

Kinda wishing I didn't afterward.

I saw holes in his body. Puckered bullet holes were seen on his baby blue button-down. Filling his chest and torso and staining his shirt with red, leaking down his pants. I started counting the punctures, but I ran off and tried not to vomit. I avoided looking at the body again when I reached *seven*.

That man didn't just die. He was killed. Shot down. Seeing death was okay on my part, but this. This was a murder. A massacre of one man? *Why?*

I walked away from the body and looked inside the busted Civic. Nothing but trash. Whoever killed him scavenged everything valuable off of him. Not even the spare tire in the trunk was there.

I wanted to break down. Give up and go home. But of course, I couldn't. On other accounts, one being that there was no such thing as *home* anymore.

I left the man and his car where it was and jumped back inside the Humvee. The engine kicked itself up again after I gave it the keys. Luckily enough, gas was still good to last. So… well… I was on the road again.

Find someone? *Check*

Find someone *alive?* Still working on that.

I continued for another 20 minutes and stopped when I saw a standing radio tower. One thing I know about me is I do know how to climb well. Very well. One of my main skills includes urban parkour, yes, the illegal kind. I used to climb fancy cranes and abandoned buildings when I had the chance. I am not a *fully* good influence, but the views at night were so worth it. But there are no laws now.

I shut off the engine of the truck and began to get out. If I could climb it, I would get a vantage point and be able to see if there was any life around. Smoke trails, cars, anything.

I paused when I heard another engine. Yeah, another one.

I looked over the door to see a black jeep in the distance with blaring cop lights on. It was moving my way fast.

?????? ?? ????

10:57 AM

Capital interstate road. Texas

"Hey! Hey!! Help!!"

I jumped and waved my arms around. Finally, people! I could almost pop.

"Hello!?" I yelled again.

It didn't slow down. In fact it looked like it sped up. When it did, I spotted another vehicle behind it. A truck. It had similar red and blue lights flashing on the grill, but seemingly, in the bed of the truck was something large and gray.

Took a few more seconds before I realized what it was. Have you ever heard of the "fight or flight" response that all living things have? Well, mine kicked into hyperdrive when I noticed a mounted machine gun in the back of that truck, which consequently switched the lead with the jeep. And a man standing in the bed, pointing it right in my direction.

I ran around and hid behind my jeep on my left side. And I did it fast enough. The popping sounds of the gun came. It wasn't like popcorn pops. Or bags of air pops. These pops sounded like a small explosion every time. I could feel the gun hitting the Humvee.

Tack

Tack

Tack

The crossfire seemingly never ended. It finally stopped driving about 30 feet away, but it didn't stop shooting. I grabbed my 1911 and pulled it up to my chest. Itching the trigger for dear life.

Finally, after a million shots at my poor Humvee, the window finally gave in and shattered. The bullets stopped, and with the open door, I could hear the radio going off, a manic laugh going off from it.

"Look here, boys! We have some tough meat!"

The clanky radio called out with a loud, Dangerous voice. As manic sounding as a wild boar, as if a talking bear was on the other end.

Still crouching, I made my way around to the driver's side with the door open and grabbed the square clunky microphone from inside, pressing the gray button with sweaty hands.

"Cease fire! Cease fire! I'm not bearing arms! Cease fire!" I begged to the crazed voice.

"This slab of meat sounds fresh! What be yer name, boy!? And your age!?"

"Will... Will Light, Sir... and I'm 17."

Why didn't I give a fake name? Because I didn't have the clarity to *think* of one.

"We have a Runt here!" It said, choking up another laugh.

"Now listen to me very closely, or my guy here will blow so many holes in you that we will be able to read a newspaper through you!!!" He yelled.

"Walk around the car and get on your knees with your hands on your head!"

Humvee, I thought to myself as I threw my gun into the car and got out standing up. I walked around slowly and got on my knees about four feet in front of my bumper, facing them. Yeah, it could be an active trap to lure me in, but I literally had no other choice. All of the jeep's doors flung open, and six men came out walking towards me. Armed.

"Don't be funny, kid!" the police intercom blared in a robotic tone.

My first thought was to fight them, but who was I kidding?! Six men with guns?! Hell no. Sure, I was okay at hand-to-hand combat. I did MMA for a few years but I don't think even John Wick could handle that. Maybe if I was in a movie, I could. But not there.

As they walked towards me, I could see they carried different types of guns. One carried a submachine gun, four carried AK-47 and one carried a Glock 19 with a barrel attachment. However, they were dressed very poorly. Casual streetwear, cargo pants, ripped-up shirts, tank tops, and only one guy with a helmet. Not even a military one, but a cop one.

That's smart. I wanted to snort. But I didn't.

They got up to me, and while four looted my Humvee, the man with the helmet pointed his muzzle of the AK at me.

"Pockets."

I emptied out my pockets as the four goons finished searching my Humvee, pulling out the bag and the 1911. They threw it down on the ground in front of me as I added my wallet, knife, and house key, which broke off from the wallet inside my pocket. They didn't take anything yet, and I just realized why.

The police helmet goon pulled out a walkie-talkie and brought it to his mouth.

"What now, boss?" he said quickly, like a top-tier actor.

I had to act fast. All I had to do was grab the muzzle, point it at the man next to him, and shoot.

Shit. I can't do that. This is the end of the road. They are going to kill me, jack my stuff and my ride, and leave me here to rot where I will never be able to reach Alice and my brother.

Fighting back tears. I listened for an answer.

"Floor him!" The voice yelled, laughing.

I grabbed the barrel and tried to fling it up, until I got hit with the butt of the rifle, which knocked me down.

This is it. Goodbye.

?????? ?? ????

11:05 AM

Capital interstate road. Texas

Over the course of the next minute, three things happened, each saving my life.

One - The truck with the turret mysteriously blew up.

Two – when that happened, I tackled the gunman to the ground, wrestling him for his gun.

Three - The others around were mowed down by bullets.

The guy I was wrestling with gave up and tried running back to the jeep, until he was shot down by more bullets. I ran to the Humvee with the AK and crouched beneath my door and seat, peeking out from around the hinges of the door. The band of killers was nothing more than a leaking bloody pool of… um, blood.

I also still heard an engine. A louder one than the jeeps. Peeking over the door, I could see a… Well. I did not know if I was mistaken, but I believe that the thing that just killed everyone and blew up the truck was a *Tank*.

The single barrel was as long as the tank itself, and a mounted 50-cal machine gun was mounted on the top of it. There were red spray paint patterns covering the body.

Did you read that correctly? A *tank*.

My day was possibly about to get a whole lot worse.

As I hid, the radio started talking again. In a different voice, a voice of iron, a heavy attitude. It was almost as if every word cost him money.

"Status report."

"Nine hostiles down. One civilian in a discharged Ex Army Humvee," said another voice.

"Is Beast with the rundowns?" The ironed voice man said.

"Negative sir."

"Alright, sir, in the Humvee. I know you're hearing these transmissions, so if you could, step out of the

Humvee and stand on top of the hood. Reply with *"Copy"* if you have gotten these transmissions and understand the procedure."

"Copy," I responded on the microphone, forcing myself to sound older even though they were going to see me in a second. I got out and climbed on top of the shot-up hood. My lungs were burning from a mix of the air and the adrenaline rush that I had. It was perhaps because looking at the bodies of the men around shows a lot more than death itself, if you get what I mean.

The tank parked at least 40 feet and I spotted another Humvee beside the tank. The only difference between mine and that one was that it was tagged with red, and it was coming closer than the tank, parking about 25 feet away.

As *gracefully* as the jeep, all the doors flung open, and instead of shirtless criminals, these were a well-equipped band of six. Four of them had bulletproof ammo-carry vests. Two with actual combat helmets with one black man having real *Infrared night vision* spectacles. The rest were wearing red-maroonish shirts and digital camo pants, different shoes too on each person, so I guessed they didn't provide shoes. Still, each of them was with an M27 Rifle in their grips, and this was the most organized unit of soldiers I had seen so far. One of the men, though, was dressed differently. He had an OCP Camo vest, and military cargo pants that matched everyone's digital design, only it was a Maroon red. It was, seemingly, no more like *obviously* the most important of them. He was definitely pushing my Sgt Parlor's height, maybe even reaching it. The differences between the two were instead of a black, balding man for Sgt Parlor, he was a white, tanned man with silver hair and a square, dirty face matching his muscular, burly body. He stood with no gear besides a gun holder for what looked to be a silver magnum.

The men around him got straight to work, looting the guns around the fallen men while the big man walked up to me.

"Off the hood, lad."

The voice was recognizable. It was that Harley Davidson voice I heard on the radio. Almost every sentence came out like a solid stone remark.

I jumped off the shot-up Humvee, and as I did, a soldier picked up a body that was next to my gear. Looting.

"Hey wait! Leave that; that's mine."

"The shorter pale soldier with no vest or helmet looks at the big man, clearly waiting for orders."

"You heard the kid. That body is his," he answered in a louder, heavier tone, making the man drop the body and my bag.

"I'll be damned, kid. I assumed you were much older from how you handled the situation. Clearly, you had practice."

I dont know what he saw. Maybe me running for my life or being cowered down by six soldiers. Maybe he was being sarcastic.

"Oh… ah, um, thank you, sir..?" I babbled like an idiot

"At ease."

Instinctively. I put myself into a position with a snap movement.

"Kid, I meant to be quiet."

"Ah, sorry, sir."

"Where are you headed, kid? Out of state?" he questioned.

"Worth, sir. I have someone waiting for me in Dallas."

"Dallas? Are you sure, Dallas?" he asked, surprised.

"Yes, to meet someone, why? Is it destroyed, too?" I asked, hoping I was insanely wrong.

"Lad," he said, going to a lower tone, oh no

"The world is destroyed. Two weeks ago, fire rained from the sky. You, me, and a few others are what remained of the estimated eight billion people that died. I supposed you should have known that."

September 6th, 2035

11:15 AM

Capital interstate road. Texas

You know that feeling when you're about to throw up, but you end up swallowing it back up?

Magnify that 1000x because after I heard that, I swear I thought my skin turned white, and last night's dinner was gonna spill any minute now.

Two Weeks?

No. He got it all wrong. It was just yesterday that everything was normal. Now I'm hearing two weeks?!

"How…" I muttered out loud.

"We intend to find out." The big man said sternly. "Come with us. We have cooked meals at our camp, along with other civilians. You look like you haven't had a meal since it started."

Hah, yeah, no, I actually had *one*, thank you.

"Cant." I choked out.

"Boy, the city is almost in ashes…"

"I dont care. I have to go as soon as possible." I stubbornly responded.

Moving around him, grabbed my stuff, packed up the little items back inside my pocket, and clipped my knife back to my hilt. And I realized I had left my lighter at the gas station. Fuck.

Clearly, my remark back to him surprised him a little. But he quickly regrouped himself and spoke back. Calm but still firm.

"Lad, it's too dangerous, me and a few men at our camp are grouping together a small convoy through Worth, Dallas, and other cities on the road. Come to base with us, and I'll seek to bring you with us."

Thinking long and hard, I looked at the bodies that were cleaned out, thinking that will help.

It didn't.

"Okay, but I'm taking my stuff and my ride." I nodded in the direction of the shot-up Humvee.

Yeah, if this Beast starts.

"Alright," he said, reading my mind as he looked at the shattered window. When I stuffed the 1911 in my pants, I saw him stick out his hand. "Colonel Dominic."

"Will Light," I responded, taking a firm handshake.

"Delta Team, Roll out!" He said in a booming voice. Almost blowing my eardrums.

"If that monster starts, dont crash into my tank."

11:23 AM

As if the spray of bullets was just a warm shower for this Beast, The Humvee started with almost no trouble.

"For a retired senior, it can take a damn hit," I said to no one as I drove behind the red Humvee with the Tank behind us. We started driving for about ten minutes before we took a turn off the interstate and into a little destroyed urban area. Two more minutes down the road, the radio went off again.

"ETA to Ranger City?"

"Five Minutes."

"Copy that."

The radio just kept chattering with random commands. Most of which I didn't understand. I noticed that this Army's color is *maroon* rather than the typical green. Maybe they weren't official soldiers?

A few minutes out, I see down the road a clearing of trees and houses. And a field of a basketball court with a track court is seen.

"Great. I'm back at square one." I groaned as we pulled up next to a *school*.

Although there were well-geared men in maroon at the entrance of the school with one having an AK and the other carrying an LMG, when we passed them I read the entrance sign of the school.

WINDOME LAKES ELEMENTARY SCHOOL!

HOME OF THE SHARKS!

BOOK FAIR NEXT FRIDAY!

I drove in behind the tagged Humvee slowly.

"Park next to me, kid," Dominic said on the radio as it echoed his voice.

He pulled into an empty, handicapped space next to the school entrance. And I pulled into a similar one next to him. Hope I dont get a ticket.

I got out and looked at the other combat units around. More Humvees and other kinds of red-tagged military engines. I shut off the Beast and grabbed my bag, strapping it to my back.

Beast… I heard that name on the radio during the firefight. Who is the Beast..?

I opened the door and stepped out. The air no longer bothered me when it hit me head-on.

"Dominic!" I said, running up to him while he walked up to the entrance doors with the barbed windows. And three geared soldiers standing by them.

"*Colonel.* Dont forget it. You want your time here then you better learn to respect my rank." He said, in a voice that was stern and firm. Not turning around, he flicked his hand up and slightly bent it. I guess it was a learned signal because the men standing by opened the doors for me and the "Colonel" to walk through.

"Welcome to the Rebel camp, boy," he said loudly and seemingly proud of the room area we were in. The only interesting thing was a cat poster on the wall talking about the book fair.

"If you can't be anywhere without an adult, then go in there." He pointed at a chicken wire window of a small room. Above it, reading "Front office."

"A lady named Ivys is inside. Her rank is First Sgt. So, think of her having as much authority over this place as me. Tell her I sent you and to send you to the civilian safeguard."

"But sir…" I began to say as I tried to restate why I was even here.

"That's an order!" he yells. Harshly loud. Surprising me a little

"Yes sir…" I say. Defeated.

I slumped myself over to the door as he walked back outside. When I opened it, I was expecting some older lady sitting at a desk. Instead, I saw a woman. Late 20's, 5'8 - 5'9. Her clear ivory skin reflected her long, black straight hair, slim figure, and pretty face with a not-so-sweet look on it. Her black eyes matched with her gray, black, and maroon jumpsuit. A gun clipped to her hilt. She seemed the opposite of me. A ratty-looking 6'0 kid with dirty blond and gray hair. If I could even name anything about me, it would only be my figure. I dont think I look attractive, a… I guess doll-like face, with bronze hunter eyes and no facial hair.

"Fall in!" She yelled as I walked in, not looking up from whatever sheet of documents she was looking at.

Like a dog, I instantly snapped myself to a standing position, hands to my sides, looking straight forward. Damn, my muscle memory for these incarcerated commands.

"Name and rank"!

"Private Light First Sgt!"

She finally looked up at me, inspecting me and my form, okay. Is it technically a lie that I did say I'm a private? I feel like Dominic is gonna go back on his deal, so I will do everything in my power to make sure it's possible for me to go if he does.

"At ease."

I snapped the movement as she looked at me, her black eyes almost steaming at me.

"Your purpose?" She asked in her cold and velvet voice.

"Domi… The Colonel sent me. He asked for me to be sent to the civilian area before the trip." I replied, making sure to sound sure.

"All right, private. Turn in your bag and gun, and if you have keys, hang them up, too. What's your locker number?"

"Um… I dont have a locker yet. Or a gun." I quivered, trying to sound perfect.

"Right. I forgot we were running low on arms." She said, pulling out a clipboard from under her desk, "Your new locker number is 140."

"Yes, ma'am."

"Bag and keys," she commanded.

I pulled out the keys from my back pocket. Feeling the 1911 that I completely forgot about. I can't give her it now. It would be a problem. So I keep my mouth shut.

I took off the bag and put both on the desk in front of her. She took it and put the key in the bag's small pocket while putting the bag on the floor.

"Fall out and follow me."

I stepped back and walked behind her as she walked out the office door and down the long indoor hall. I only trailed behind as she walked up to a soldier standing guard. I also remembered my knife, but I guess she didn't care. It was very visible on my hilt rather so. Anyways, they exchanged a few words. He was taller by about an inch, and he looked fierce like he was getting a bonus to just look intimidating. Then she turned to me.

"Follow him. Your stuff will be in your locker. We will be leaving in T Minus Fifty hours. Be prepared. Go."

"Yes, ma'am."

I hurried towards the man that already started walking down the hall. We walked down until we turned right when we reached the end and kept going. In silence, we made it to a steel door guarded by two other soldiers. Damn, this really is a base.

"You're quite young to be in this Army. How did you get in?" The man who was escorting me asked as we walked inside the steel door to another hall.

"Colonel let me in."

Okay. Technically, that's not a lie.

"You got past his physical camp? Color me impressed," He said, shuddering.

Damn, that must have been rough times.

Finally, we reached another set of metal double doors. The side card read "Cafeteria," though.

"When you get in, try not to look into the eyes of anyone, got it?"

It was a confusing thing to ask for. But okay, was everyone wanting to be treated with respect?

He opened the door, and I realized why he said that. Answering with a loud "No" to my in-thought question.

11:57 AM

The floor was littered with activity.

Not the good kind.

All around on the ground were rows of blankets and mats. IVs sprouting out at random spots. And almost every mat had a person in it. Bandaged up, bloody, some even missing limbs. And it wasn't just soldiers. I would assume more than half of every one were regular people. Some even looked younger than me.

"We are in the kitchen. Go through those doors, and you will make it to the cafeteria. AKA the civilian camp."

"All right, thanks," I said, walking over to where he pointed, merely avoiding stepping on anything or anyone.

The thin path was hard to stay on with so many nurses and maidens walking through. All of them looked more tired than the last. I made it to the door as I heard the heartbreaking chatter.

"I won't be able to see *ever* again..?"

"What do you mean my body is *burnt?*"

"I'm trying to grab it, doctor…"

I can only just wish you the best. If I could do more, I would. I grabbed the steel handle and pushed it open. Hoping I dont see more of what i seen already.

Inside the cafeteria was better. I expected roll up beds and random people just going about their day in them but instead it was a rainbow jungle of tents. It almost looked like a miniature neighborhood. Or a homeless hideout to be a little more specific. The barrels of fire really added to the touch. There were rows between every two stacked tents. And a loud chattering of murmurs and voices came from all the people rather inside them or around them.

I was confused about what to do. I had no place here and it looked like there was no room to make one. Not that I had a tent on me either way.

So i turn. Maybe I'll find somewhere outside…

"William..?" said a voice behind me as I turned around. An all charming and way too familiar voice in a way too familiar blue dress.

September 6th, 2035

12:04 AM

Ranger. Texas

"Mom! Mom!" I practically charged into her arms, slamming mine around her.

"Wha- Will… How are you… Oh, Will." She stuttered out as tears came down her cheeks. And I lost the battle fighting back on my own. Even if *seemingly* a day passed for me. I didn't care. At least I knew that Alice was safe.

But that was just half the battle.

"Will," she said, pulling back to look at my face, "I thought something bad happened to you."

"Of course, Mom. You're always right. I got knocked out for a while. I woke up and did this and that, and now I'm here!"

"Thank God," she said, crossing herself. "And God bless your brother to be okay."

Oh.

"He isn't here?"

She shook her head, Dammit.

"I never managed to reach him," she said, "But don't worry. He's in good hands, and the hospital is very sturdy."

I don't know if she was reassuring me or herself.

"I'll have to see for myself."

She looked at me with a perplexed expression. "William… you can't."

"It's my job, Mom," I said, flashing my ripped coat.

"You… you enlisted?" She asks, sounding sincerely shocked.

"Yes…" I responded, trying to sound confident as I watched her face fall. I could see her eyebrows furrowing. *Oh no, I'm about to get the life lesson of a lifetime from Mom. Again.* I thought to myself.

"No. No, no, no. You're quitting!" she said, raising her voice slightly.

"What?! Why?!"

"Nope! I'm not allowing it!" she said, followed by repetitive "No's."

"Mom, I literally have no other choice. I need to find Jack for us." I tried pleading with her.

"No. End of argument."

"Dad would want this for me! What do you think I'm trying to do? I want to bring Jack back *safe,* Mom!"

She just stared at me. Studying me. I felt horrible pulling out the dad card. I only ever did it once before, that was a few years ago. I was in another begging battle with her to see Jack, to visit just once. I lost, and when I did, I was already fuming. I yelled at her "Dad would let me!" And the expression on her face was cold. And now it's showing again, making me feel horrible all over.

"I just…" she squeezed her temples. "I dont want to lose anyone else."

"Mom…" I started to say in a soft, calmer tone. The one I use when I try to be reasonable with her whenever she is mad at me, or one that I try when I try to calm her while she is hurt.

"Mom, it's okay. I'll be mostly protected. Plus, I'll be helping us find a good home!" I said, grabbing her cold hands. "I'll get Jack home. I'll find him. Please."

Eyes flooding with tears, she nodded.

"Thank you." I closed in another tight hug. But with my bruises, I winced a little.

She heard my wince and looked under my shirt and at my neck. And she reacted as you expected, clearing up her feelings.

"Oh my god what happened to *you*?! Did you crash?! Why have you not told me?! When did this happen?!"

"Mom, Mom, Mom, calm down. You're gonna cause other people to look. It's just bruises and little cuts. They will heal. It was from a cave-in, you know, my school?"

"Let's take you to the nursery," she said, gesturing me to the door I came in from.

"I'm okay, I promise! I… just need some ice. You know I heal quickly, Mom," I told her, trying to avoid the depressing feeling growing in the room.

"Are you sure..??" she asked. Worried, of course.

"Let's go to your tent!" I said quickly

"Oh, right. I guess it's time for them to serve lunch."

"It is?" I ask, taking out my watch as it ticks to 12:30. Woah. It is. That was fast.

"Are you hungry?"

"Yes!" I responded before she could even finish. "Ahem, yes," I said again with a more moral sense.

She smiled. It felt like forever since I saw that heartwarming smile. Even during the worst of times, it's amazingly powerful. It almost resonates with her.

We walked back to her tent. Once inside the neon orange tent, she started questioning me like a cop— asking where I'd been, if I was okay (she must have asked that one at least twenty times), how I'd join the rebels, yada yada yada. After answering her as best as I could while we ate in the tent, I turned my attention to something on the menu that felt odd— a mysterious soup. It tasted fine. It just looked weird. Anyway, after we ate, I had questions of my own to ask her. And oddly enough, I did find answers to a few things.

The army that I seemingly joined, the one that was protecting us, was named the *First Rebel Independent Ruins Army.* It is a PMC group made up of ex-solders from the army or other militaries, all under the supervision and command of two commanders who I already met. I asked what happened to the actual army and government. All she said was, "They will come around." So, from that, I guessed they are trying to figure things out, too, if they are out there, that is. I asked how long she had been here and she answered a week and five days. Of course, I asked how she was and how she was feeling, which was a mistake on my part because it reminded her about my injuries. So I was stuck laying down while my mother was putting ice bags on me that she got from the nursery. It was like that for a couple of hours until nine hit. Then, it was lights out for everyone. All the soldiers had a few classrooms as rooms, so there wasn't much of a mess of civilians and soldiers sleeping together. Only a few did, and the ones who did had to keep their tent doors open. I didn't have an assigned room with any of the soldier groupings, but at least I was with my mom.

If she felt lonely, at least she was lonely with me, making it the two of us.

September 7th, 2035

4:00 AM

Ranger. Texas

I didn't sleep until the morning hour like I was usually used to. I was awakened by a man with a flashlight on my face.

"*Up, soldier.* We have a morning announcement. Get ready and go to the track field," he whispered.

Ugh. Being a soldier sucks already.

I pulled out my watch and very intently looked at the face. The glow that was there in the dark really sucked from what it advertised on the website, merely making out a *4:02* dial that it read. Definitely not close to the five-thirty I was used to.

Quietly, I climbed out of the tent and slid on my boots, which were off and next to the tent nicely together. I forgot I even did that. When I got out of the tent, I looked over at Alice, and I had to leave something of mine behind to ease her morning worry. The wallet, knife, and gun were all that I had so being the Oh-So-Smart son of a lifetime. I took out the watch from my wallet and set it on the tent side that I slept on. It may help her tell the time by not looking at the sky because there was…like no sun. None at all.

With that, I walked out and followed a couple of other soldiers outside, where all of them were sitting on the bleachers in front of a marked field of an oval. Everyone was scattered all around, and there were *lots* of people. I didn't know there was *this* many soldiers. I could count over a hundred here now. And lucky enough, I didn't seem to be the only one who was a little confused. Recognizing the confused scowl on a few people, I felt more relaxed. Quiet whispers and words were exchanged between soldiers. Luckily, there are no bugs out. Or possibly even *alive.*

"At ease!!!"

The sudden yell of the colonel and first sergeant together made me jump. The people on the bleachers clearly were also surprised due to the conversations turning silent so quickly. Not even the morning birds made a sound — oh wait. Forgot.

"Thank you, everybody, for coming out this early!" he started as if we had a choice.

"As you all may be wondering, we have not yet seen any official and active response from the United States government or any of its branches. Announcements, plans, and military responses are still yet to be found. So as of this moment forward, we will treat ourselves as if we are on our own."

Great. So, we are *officially* alone.

"I know almost everyone here has lost someone from the event and some from the events that followed. As further facts conclude, we may be one of the very few small defense forces that formed after the meteor shower."

Low whispers and murmurs started again from the crowd.

"To anybody who was or is confused, the recap goes as follows. Approximately 15 days ago, meteors entered our planet's atmosphere. Fire and ash covered not just the atmosphere of America but the world as we know it. Amidst the clouds above, the sun and moon still linger, just covered. 12 days in, and almost all known vegetation or plants have rather died or entered a dormant state to survive. 14 days in, and most of all, species of life have had an effect, too.

I was just reminded of the wolf I saved in the forest. Was that a wasted effort? How many people could even be alive out here? How many were left?

Could the wolf have saved me?

"What about the sentinels?!" A worried voice shot from the crowd.

"Yeah! and Beast?!" yelled another. As the voices from the crowd voices began to grow I started to think once again. I heard the name on the radio when Dominic saved my ass. Who is that? Or what is that?

"At ease!" Dominic barked in a tone surprisingly loud — loud enough to silence the hundred eager voices.

"As you all seem bothered for, The sentinel army grows in number to this day. However, they will be dealt with as swiftly as possible. With you helping us deal with this opposing force, the criminals that had been gathered will barely show any mercy to the innocent lives of most. That's why we need to work on finding their weak points and finding reasons for prosecution in their actions.

"Prosecute?! They killed my wife!"

"They deserve to be hanged!"

"They will burn in Hell!"

A riot of angry and stricken shouts boiled through the steaming crowd. It became apparent that nobody was happy with these "Sentinels." And whoever that Beast person was. Beast seemed to be another issue.

"SILENCE!!!"

I never even guessed how loud a man's voice could be. Instantly, all the yelling stopped, and his voice could be heard echoing in the distance.

"As I was saying!" He continued. Quieter but still very loud. "Beast and his army will be dealt with. As for now, our worry relies on our own lives," he said slowly as if he was talking to a room of children who didn't understand a concept of something.

"Speaking of such, that's the reason I have brought you all out here!"

September 7th, 2035

4:24 AM

Ranger. Texas

"Our mission consists of three key elements! The first being hospitality, all civilians that can, and that are needed to be rescued, will be housed and protected by the rebels."

"Second," Dominic shouted, sticking out two gloved fingers, "We will be finding a home for us to sustain safety and life to everyone here and everyone that comes with us, for if we are the only ones out there, that is," he quickly added on.

"And third," he continued, holding up one more finger. "Maraud, the military bases for supplies and Pharmacopoeia for us. While doing so we will be traveling a path to three specific military bases to show in code 797. A code set with the purpose to find and secure all help from any government and or armed forces bases in the country," he said as he put down his hand. "When we do get to these bases, we will find the contrary components to activate a powerful SOS signal. Therefore, all remaining civilians can get word and come to us."

The murmuring began again, and my mind was racked with a lot of things, but mainly with something that someone from the crowd spoke of.

"Um, sir?" a female raised her hand from the crowd — a small figure, redhead. "Where will you go, and when will you return?"

Notice how she said, "You?"

"Our path leads to Worth and Dallas first. After our gathering, we will head to Houston. Both destinations have large enough military bases to have the code 797 machine. We already activated one, so all we need are those two. We will also pass more cities along the way, checking all of them. We may not be back for a while. A month, maybe, and even with whoever comes, there is the risk of you not coming back," he finished, igniting harsher whispers.

"Here are my three promises to you if you do choose to come or to stay. You won't be punished or held as weak for whatever action you choose. Bear that in mind."

"If you choose to come, you will be treated as you are here."

"And if you are with us and trouble stirs, if a battlefield if one is ever framed for us," he said, pausing, "I promise under my name and under my rank that I will be there for your protection if you are there for mine. Every soldier that comes with us is a link in our chain. A gear in our machine, along with me and Ivys," he said as he stuck his hand out to the woman that I saw in that small chicken-wired office before.

The crowd was silent other than a few quiet whispers that were still stirred up as Dominic continued to speak.

"We will be leaving tomorrow at 0-1200 hours. Anyone who chooses to come will meet Sgt. Ivys at the concession stand by the exit gate. I hope you all seek through. Fall out!"

As I saw the large crowd get up from the bleachers, I could see hundreds of heads and thousands of bleacher stomps filling the air. All of this made me think that we would have plenty of soldiers to accompany us.

September 7th, 2035

5:40 AM

Ranger. Texas

Boy, was I wrong.

Out of the 500, only 40 men and women lined up together. I guess no mention of a paycheck caused some heads to turn away. Or it could be the mention of danger that caused people to scurry out without looking back. The conversations between them were pretty loud and obvious to pick up.

"That crazy old fool is a sellout! I got nothing from the first mission!" A man with a ruffled voice yelled in between people from the crowd going back inside, confusing me. Like, money was worthless now.

Yeah, it's stupid to stand in a line of soldiers ready to deploy into possibly active combat. But risking my life for my brothers is a deal I would take any day, and this is that deal.

I lined up with the other 40 or so that are risking themselves. Why? They have their own understandable reason. Like I do —at least, I hope I did.

"Next!" Ivys yelled after about half an hour with the sun starting to show its golden hour.

Oh yeah, my cue.

I stepped up to the camo overhead cover like I was stepping up to the Grim Reaper himself as he figured out what he was going to do with me.

"Name?" she asked, tapping her pen onto the clipboard she was holding.

"Will Light First Sgt."

She started writing, "Locker?"

"One forty."

"Are you driving, or do you need to be issued a seat?" she said, not looking up from her clipboard.

"I have a Humvee, ma'am"

"Is it tagged?"

"Tagged?" I asked.

"That's a no. Tag your car to a noticeable trait. There is some red spray paint over there," she said, pointing the end of her pen to a box on the floor beside her. "The only rule is *Tag*. Dont write anything stupid. Your keys are in your locker. Along with issued supplies," she says, finishing with her scribbles. "Next!"

As I stepped out of the line, I opened the box and saw a bunch of red spray paint caps on the cans. I take two and walk with the ends of the crowd that were walking back inside. The good news is that there will be a lot of soldiers around for Alice.

I walked into the doors where I came out, passed by the cafeteria, and made it to the hall where people were all heading in and out of classrooms. Again, it was a mix of soldiers and civilians. However, it seemed school-like because of the chattering and the clash of soldiers going into their lockers. And I found mine, the silver box matching with everyone else's. It was about four feet tall so I bent down to dust off the number plate on it. Opening it, however, left me stunned.

I didn't really know what to expect opening the locker for the first time, but definitely, not too much more than what I already had was not one thing in my mind. My keys were in there connected to my bag, but just looking inside showed me a new pair of cargo pants and a V-Neck tagless maroon shirt I saw some of the other soldiers wearing. I put the paint cans down and started digging inside like a kid in a candy shop. I pulled out the clothing, as well as military class boots, A small flashlight, and five MREs. (Wow)

Thought that was all? Don't worry, I saved the best for last.

Inside a holster that was meant to go on my thigh was a serious spy-looking pistol. I looked at the steel slide as it read in perfect engraved text "SIG Square M18." I'm a dumbo when it comes to some guns, but for this one, I understood that "SIG" means it came from the Swiss German company. They really handed a gun to a kid while at school. God bless America, am I right?

Besides the clunkiness, it looked brand new. And with four mags, it could do some damage. But that's nothing compared to my last item.

I pulled out the last steel-cladded item. A dangerous *M27* lingered in my arm.

6:25 AM

With the heavy steel of the American-made gun, the weight clearly showed its destructive power. The cold metal on my shoulder also felt as if… weirdly… it belonged there.

Well, not in the middle of an elementary school in front of everyone here, but still.

But the feel of the rubber handguard and the pistol grip made me feel… bigger.

I dont know. I may just be psychotic or weird.

Or both.

I held the gun by its muzzle and put it back inside the locker. Hoping not to get too many stares. There was also a padlock inside with a key in the corner. I didn't see it at first, but now I do. I packed everything back in my bag, including the extra two filled mags of the M27, and put the backbreaking bag inside the locker, taking my keys. I closed the locker and stuck my new padlock on it. Sliding the key inside my pocket as I picked up one spray can. I put the other one inside, just in case of anything.

I think you can figure out what happened next. Went outside, passed the door guards, and went to my Humvee, which (lucky enough) was still parked in the same place. It took me about two hours to do *half* of what I planned. Nothing stupid, I know, but I ran out of spray paint. So I'll have to get more if I do want to finish the design. I go back to the doors and walk inside.

Well. I tried to walk inside.

"Woah woah, hold it, bud!" said one of the guards.

"Is there a problem?" I shot back.

"Yeah, I've never seen you here," he sneered.

The man was Asian, and he was surprisingly as tall as me. He held an M16 on his side, and despite his very white skin, he was quite muscular. I guess all the gear he was wearing yesterday hid it because today he only wore the signature maroon shirt that his arms stretched out the sleeves of. He wore cargo pants, too, that matched with every other soldier, including me.

"You saw me yesterday! Look, I have the same issued cargo pants you have!" I pointed out, annoyed.

"First of all, you were wearing those yesterday, second of all. I said I have never seen you before. Nothing else," he said in a fake sweet tone. I imagine this is a tone a wolf would use to lure in a hare.

"Lovely. Now, if you will, please excuse me-" I pushed past him and pushed open the double doors.

"In a rush to say bye-bye to mommy and daddy shrimp?"

I stopped at the doorway when I heard that — briefly.

"That is if you're going with the big boys," he chuckled with his goons.

My father told me before to choose my battles wisely. And so I did.

"Ah, right- you just reminded me. Thanks," I said, walking off as the door slowly closed. I did hear him curse something, but I chose to ignore it. Such a nice guy. Maybe I should share tea and cakes with him.

I pushed that situation past me and, got back to my dusty silver locker, and opened it. Grabbing everything besides the M27, I walked to the bathroom down the hall.

I never knew a school bathroom could look worse than the one at my academy, but with this one's trash litter of several thousand people it really does take the cake.

I locked myself into the biggest stall and kicked away some silver cans and wrappers. I set my heavy bag down in the clearing and began to sort it out.

Listing everything down that was counted would take way too much of this page. So I'll just tell you what I packed and what I planned to leave behind. The list coming with me included:

- One flare
- One trauma kit
- One compass
- Five MREs (Still like. wow)
- One canteen
- Extra ammo for the guns

And that was all the normal supplies. The envelopes stayed where they needed to and I strapped the thigh holster on with the gun inside. It was a new home for the next... whenever.

I rechecked the knife and felt a slight pinch on my lower back. I pulled out the object again, and I was conflicted when I felt the 1911. Pulling it out and inspecting it, I realized I couldn't take this with me. I didn't have the right ammo for it anyway, and it's retired. Plus, it's my dad's. It's not really mine to have. I'm sure Mom would be a better owner of the gun.

I grabbed my bag from the messy floor and zipped it up. Sticking it on my back, it felt way lighter. I guess all those canned foods and pots really added up weight. However, all the rest was going to Alice. I quickly found a large plastic disposable bag from the trash-filled floor and stuck everything I didn't take with me inside. Aside from the canned food and the gun, it was all stuff she wouldn't really need unless it was a certain situation. Good to have, though.

This was the easy part.

Now for the hard... no— more like the challenging part. Actually, really, really, really impossible job.

September 7th, 2035

3:07 PM

Ranger. Texas

Later on that day, I was eating outside with Alice at an aluminum table, among a few others. We ate the lunch that the camp handed out in peace, though I was about to change that. Rice and chicken weren't really my choice for a last meal —at least not if it was the last meal of my life.

"You know you do look like your father," she said while eating.

All I could do was nod with my mouth full.

"It's going to be funny seeing you marching around," she continued, snickering.

Usually, I would blush, embarrassed. But I won't have to worry about that. Maybe.

"About that…" I began to speak as I swallowed my food. "Mom…"

She read my expression. Her face instantly turned into the familiar worried side I'm used to. "Scared?" she asked, getting the wrong idea.

"Oh… yeah… I mean, no. Wait, yeah… but not for the reason you're thinking," I tried to speak, choking on not my food but my words.

"You know it's okay," she said calmly. "I'm sure the jobs you do will be easy. No need for stress!"

"About that again… I kinda already have a mission."

"Ah," was all she said. "Well, that's… good," she said, coughing at the words that were to come next. "I'm guessing you're just nervous."

"Yeah."

"When will you be back? Few days, perhaps?" she asked, lowering her plastic fork.

Good. she can't stab me yet.

"Well. We are going to Worth and Dallas. After that, we will head to Houston. We will look to find ourselves a more reliable home, at least more reliable than a school… so we will be checking every city on our path. So our estimated time is… a month.”

“A what?!” she yelled. “A month?!” she threw back. Standing up, “You're not going! No way in hell! You're staying here!” she said.

“Mom I…”

“Nope! I dont want to hear it!” she shouted, stalking off back inside the school. I'm guessing back to the tent.

I had no idea how I was going to get past this.

I took my trash, but I left Alice's plastic plate. Her rice wasn't finished, so maybe she would come back for it. I leave it behind and headed inside. It was *3:15,* so I had to find something to do while I waited for tomorrow. Or at least while I waited for the night. Clearly, Mom was not in the mood anymore to even see me. I guess I should just bring all the stuff I left behind inside the locker and bring it to her tent. I'll see if I can make her something extra special so she calms down.

And so I did. I made it to the locker and unlocked it. Before I saw the disposable gray bag of stuff, I saw the second red spray paint can. I forgot all about it, so I guess that was a good time to finish tagging.

I grabbed the can and locked up the locker. I did end up passing one of the door guards from earlier, recognizing his red beard as I made it to the exit doors. Walking up to it and hoping I would not see Monroe.

Kudos to my prayers. As I walked outside, I saw him with one other door guard. He was undergeared and didn't even have his *shirt* on, showing his insane build. He was smoking a cigarette when he spotted me.

“Hey, Shrimp!” he said, dropping the cancer stick and stepping on it. “Whatcha got there? Paint? Hey, while you're out there, don't tag “HELP” on the road, please,” he said, roaring with laughter.

All bronze and no brains. Anyway, I just walked past him and headed to my Humvee, which was… already vandalized with some other red spray paint. The wolf design I was almost done with was sprayed all over with more red paint, like a toddler using a crayon for the first time.

It didn't take an Elon to know who did it, especially with that soldier sharing glances with the other soldier. They seemingly all started snickering as I looked back at them.

Seriously, what's up with him? We are on the same team, are we not? I let my messy gray-blond hair make a tarp between him and me. At least this tangy hair will serve some help.

So, I started painting. Working with the random scribbles, I made a wolf with a large line through it. Though I had a feeling you wouldn't recognize it if I wasn't to tell you what it was.

Though I was choosing my battle, I walked back inside, and they said nothing. So I said nothing, either. I made it to the hall with no issues.

I spent the next five hours just… reading. I got distracted here and there because there was the raid drill and the formation for the pledge of allegiance with some afternoon announcements. But I made it to the part where Katniss Everdeen got put inside the arena. And she almost got a knife to her heart. I had to stop when I spotted Alice walking past the tent we were supposed to be in. I heard soldiers yelling for everyone to come inside at *8:00,* but Alice was still not going to the tent. She was here. Just walking around. I guess it was my time to shine.

What would Dad do?

What *did* Dad do?

May 7th, 2035

8:04 PM

Ranger. Texas

"Mom, can we go outside?"

"If this is about the mission, then you're talking to a wall."

"I would rather we talk *outside* than disturb anyone around."

She paused for a minute, looking around at the tents of all the people. Deciding.

"They said no one outside," she said after an eternity.

"It's fine. You're with me." I reassured her.

After about a minute, she nodded. Just once.

8:27 PM

"I just dont understand what's gotten into you!" she said for the thirteenth time. After talking for about twenty minutes, we had gotten absolutely nowhere.

"I dont know. Maybe my mindset consists of worry from half the world. And *half,* my brother. Worry is what has gotten into me. Into us, Alice," I almost snapped at her, hoping she could get it if I used her actual name.

"So instead of keeping a guaranteed safety here, you choose not only to enlist but put yourself in the field of battle for months at a time?!" she said, spitting at the word "Time."

"I chose to join because Jack is out there!" I pointed to the open field of dead trees, grass, meteors, and destruction that was barely visible under the starless night. "Do you see that? Do you think he will be fine out *there?!*

"No! That doesn't-"

"So why won't you let me go get him?! Or at least find him!" I almost yelled at this point, cutting her off.

"I dont think you can handle what's out there any more than Jack can! This is not the army with countermeasures and organization! This is an *apocalypse!*" she yelled. Very loudly.

"Exactly my point," I tried to reason with her, lighter, calmer in tone. "No one's out here to find Jack for us. No one is out here to fix this mess, Mom. It's just us. So if we don't look for him, no one will."

She paused for another moment. And I swear I'd never even seen this level of shock on her face.

"Mom. *Please.* I know where my brother is. I went into his room-"

"WHAT?!" she yelled— again. And at that moment I would be surprised if anyone *didn't* wake up from that.

"Listen," I said in the same tone that I kept calm. "I went into his room after I found the key to it. No, I didn't look for the key. I just found it. Anyways, I went in to see if I could pack any supplies. Instead, I found three envelopes." I said, grabbing them from the small zipper of my bag on my back as her eyes fell to the floor.

"Two from Thomas. And one from the hospital…"

"William…"

She started to get teary-eyed. Something I never wanted to cause nor ever did in the past. I haven't seen her cry many times. I could only count three. The one I vividly remember was the soldier at our door telling us the bad news.

"Mom," I walked up to her and hugged her. Unsure to do anything else. She sobbed gently on my shoulder… and heavily on my heart.

"It's okay, Mom. I really wanted to just talk about it. I dont care that you hid the letters. You have your reasons. I just wanted to keep the rest of us safe, but if you dont want me to go, then I won't go-"

"No," she cut off as she hugged back. Her speech faltering. "I… I'm scared, Will. I'm just scared for you.

All honesty was proven by her reactions. And I already knew she was telling the truth before she even said it.

"I am too, Mom," I told her, fighting tears again. "But I'm more scared for Jack. I have to go find him. Or at least look for him."

She pulled away and dried off her face. "You have grown up to be just like your father, Will…" she said, her tears subsiding and her sniffles fading. "Just… Promise me you will come home! Okay?" she pleaded, hugging me once more. "Promise me."

"I will," I promised, hoping I can keep it.

We went back inside. I swore I heard someone listening in, but in reality, with how loud we were, there was no one who couldn't hear that. But I still hoped that no one actually *did*. Anyways, I told her my locker number and what I left inside it (skipping the part where the gun comes in). After I did, we were already inside the tent. When I packed down myself, taking off my boots and bag, she said she wanted to give me something.

"Here, William," she handed me a necklace. With a solid, thin, stainless steel coil of a chain link, the pendant didn't match. It was a clear vial with gold caps. Inside, there was dirt and… a seed of some sort.

"You will understand it soon enough. Just keep it with you," she said, closing my hand around it.

"I promise," I said, putting it over my head and onto my neck.

We turned off our last little flashlight that lit up the tent, and after not too long, I checked my watch one more time. The face showing a *9:30* PM on the dot. I drifted off into a dreamless night, praying that the whole argument was worth it.

September 8th, 2035

8:30 AM

Ranger. Texas

"Williammm…!!" was the first thing I heard when I woke up the next morning. Everything kicked into my head when I opened my eyes at the roof of the tent.

"Huh..?" I trudged, droopy as a do-do bird.

"It's thirty past eight. And soldiers are being told to be ready by thirty past eleven. Your mission, Will," said Mom. Her face was smiling, but it seemed like a painted smile.

I groaned myself to sit up. This time, no stars filled my head. Though I groan at that, too.

"Good morning to you, too," she cracked a smile. "They served eggs a little bit ago. I saved you a plate."

"Mornin' Mom. And thank you," I said. My stomach grumbled when I grabbed the paper plate, causing her to laugh slightly. Though I was entirely grateful, I slept in. If she didn't wake me, she could have easily made me miss my trip. I owe that.

"If you want more, then please ask. I may not see you for a long time, so stuff up."

I'm glad I could ask for more. But it was hard to keep my composure to not scuff it completely down. Eggs are the best, especially if they are cooked right. It's not *technically* a home-cooked meal, but it definitely tasted like it. After finishing, I handed my plate over, asking her for another.

"Did you swallow it, Will? Jeez. Don't choke on the plate." She says. Taking my plate and getting up, walking out. I guess I should get a little ready. So, as she's out, I put on my steel-toe boots and throw on my bag. I slept with the thigh holster that I had completely forgotten about. It still was attached to my leg. And weirdly, Alice didn't question it.

She walked back inside with two plates filled with scrambled eggs. She handed me one as I took it and sat, crossing my legs. She smiled as I choked on some. Not for anything, but it had been a while since a morning started with a genuine smile from her. Not even the end of the world was keeping it down more than it usually was. Because while the world was not… demolished, I haven't seen a genuine grin. She looked tired and aged, but it looked like an actual smile.

The worst really brings out the best out of you.

"Are you ready?" Alice asked, finishing her meal. And surprisingly, not with coffee. I wonder how she will go through the day. Maybe with all that was happening, she was too distracted to be cranky.

"Never was, never will be. That's what makes it fun," I tell her with a sweet act, practically batting my eyelashes like a Disney princess.

"Such a poet," she sighed, taking my plate.

"Yeah," I laughed out.

Don't get me wrong. I like poems and stories I love more. I'm kind of a nerd in that stuff. I love making up short stories of heroes and villains. Sounds childlike, but as I said, I never had much time to fall in love with video games or something down that line.

"Did you manage to look inside the locker?" I asked, sitting back.

"You never gave me the key," she answered, leaving me to slap my own head.

"Right, here," I pulled out the right key and hand it to her. The announcement bell rang as soon as I did.

"AVF Soldiers. Please meet at the football field in two hours. And state your parting words to any who are of concern to you."

9:50 AM

I was inside the bathroom washing out my face, waking myself up. Already, we were two hours ahead of schedule. And I wasn't the only one doing last-minute preps as other soldiers came in and did the same. Slams of lockers were heard outside, and footsteps of heavy boots followed.

This was it.

I got out and made my way down the hallway through the familiar double doors that we went out through last morning for the announcement. When I saw my mom standing, she changed out of her blue dress and was now sporting a suit shirt and brown pants. She stood next to the door with her hands behind her as if she were mourning at a funeral. I hope she was not practicing.

"Ready?" she asked.

"No."

"Good," she answered. "I would have called you crazy if you said yes."

We walked out of the entrance, and on the field, there were soldiers. A small amount lined up down the dead grass and into the crosswalk. From which the camo tent, instead of being by the exit gate, was in the middle of the field.

"Drivers go on the right side," she said as she walked me to the end of the crosswalk.

"Thank you," I said to her, turning.

"Wait. Will…"

I turned around again as she grabbed me in yet another bear hug. "Dont die."

Wow, that was the heaviest emotional response I have ever heard from her. Grim, but it was something.

"Relax Mom," I responded, hugging her back. "I'm getting Jack."

"I love you, William."

"I love you too, mom."

I pulled away slightly because I knew she was never gonna let go if I didn't. "Gotta go, Mom."

"Good luck," she whispered as I parted ways with her and turned to the right line.

September 8th, 2035

10:00 AM

Ranger. Texas

After a few puzzled looks and some laughs, I stood in front of the tent on the driver's side. Dominic was marking the drivers and Ivys the riders. Though he was slightly surprised when he saw me with an M27 on my back, he wrote my name and told me to stand with the rest of the soldiers, which I did.

And twenty minutes later, when our mosh pit of forty two men was together, Dominic went onto the first level of the bleachers and treated it as a stage.

"AT EASE!!!"

Silence filled the air instantly.

"Congratulations! You all are the stronger-minded, stronger-willed soldiers. All ten of you!"

I looked at the forty-two soldiers around. Remarking his words. After a quick count,, I revised that... I dunno, but I felt like we gained two more soldiers rather than lost *thirty*. But that's just me.

"Thirty of these soldiers are actual boot camp roaches of mine who worked with the rebels before the rain of fire. The rest of you are not. However, I am happy with this team. Soldier or civilian. I am still proud to have you on this team.

That's a huge surprise. Out of the *hundreds* that were lined yesterday, only not even forty-two but *twelve* soldiers came from that. The sentinel army must really be a scary force. Or does everyone else hate free labor?"

Two very valid reasons. And two, I choose to ignore.

"We will have two drivers in our carrier trucks. Each one can carry fourteen to sixteen passengers. Including the drivers, our tank can have three personnel inside. And we have three Humvees, so we can carry four soldiers in each. When you're assigned a Humvee, you will go to it, and all the rest of the soldiers will sit in the carrier trucks."

I looked around at the huge, burly men, all with faces eager to ask questions.

"We have six drivers with us." He continued, "The two I call will be driving the trucks. Kazi, and white."

A man and a woman walked to the front of the mosh. The man was, of course, firm in muscle. Though comparing him to the Colonel made him look *small*. His dark, tanned stature reached only to his nose. While the woman, smaller, was still in good shape. Her dark skin barely glistering. Maybe from her perfect tone of skin or the fact that we have no sun.

"Tank drivers! You know who you are, you too blaze!"

Three other men walked in a separate pile in the front of the posse.

"Finally, the humvee drivers! Simon! Find First Sgt. She has the four soldiers you will be taking. Ijin, you too!"

Two more men walked to the front. One I recognized as the red-bearded man; he was shorter than me, but he was definitely way more muscular. It seemed I was the runt of the males here.

"Light, you will be escorting me and Sgt Ivys, along with one other soldier. One of our second lieutenants."

I walked up to the front as I heard whispers behind me, all lasering me down. I saw Dominic point his hand at a man I barely knew and already hated.

"His name is Monroe. You will be happy to meet him," the Colonel said, more like — a command.

"Certainly, sir."

10:46 AM

After we sorted out our groups, we headed out to our vehicles. Of course whoever wasn't called was commanded to pick a seat in a carrier truck.

"Light, you went a little overboard with the tagging," Ivys said, staring at my messy red children's artwork.

"Right. Too much. I realized that. I'm sorry, ma'am."

Choosing my battles wisely.

I stepped into my driver's seat and planted my bag in the center of the front. Monroe did the same. His bag was more or less like mine. Bigger, and it had a regular wavy camo. And he was clearly as happy as a clown at a funeral. Ivys and Dominic sat in the two seats closest to the doors behind us.

The trucks drove out first. The tank followed after them.

"Follow the convoy. You're behind the first Humvee," Ivys said.

I drove onto the street outside of the school behind one of the Humvees. The second one was behind it. I could barely make out the Ijin figure in the driver's seat. However, everything in front was mostly visible due to the fact that my windshield was literally *nonexistent.*

I already said my goodbyes to the base — and Mom. I wished them both luck as I hoped I could keep my promise.

`"Radio check, Echo team!"` the radio yelled, making me jump in, which Monroe was clearly entertained by.

`"Cargo one,check."`

`"Cargo two,check."`

`"Artillery one,check."`

`"Cruiser one,check."`

That was my cue. I grabbed the microphone and held it up. The different voices made it obvious that it was my turn, but before I said anything, a big gloved hand took it, and the Colonel did his job.

"Cruiser two and three, check. Everyone, roll out and follow the car in front of you. Cargo one, start driving to Dallas."

The roar of the first engines moving started, and I was beginning to question why Dominic even wanted to be inside the one Humvee with no windshield.

Goodbye, civilian life.

September 8th, 2035

12:30 PM

Capital interstate road. Texas

We left Ranger about an hour and a half ago and the trip was going smoothly enough. Aside from our tank getting stuck for half an hour, it was smooth. We had another two to three hours before we arrived. Why three hours? Because we were moving so slow. Radio silence was active, so it was very quiet. Boring to be clear. With so many people I never felt so lonely. I barely heard breathing from anyone in the Humvee.

Maybe that was because of the window, but it still spooked me.

2:29 PM

Stuck again. This time one of the cargo trucks carrying soldiers.

5:01 PM

Did I ever get to mention that likeubble. It weighed hundreds and hundreds of pounds. So imagine my arms after carrying numerous pieces of it off a street just so we can get past a fallen building.

7:00 PM

Another two hours passed, and the light was starting to go down. It wasn't a shock to anyone that we were all tired. But good news arose.

"Colonel, we are fifteen miles away," the radio said.

Dominic got up from behind me and grabbed the microphone. "Radio check, echo team!"

After the check, no one seemed to be lost at least for now.

"Park men, it's getting late and we need our full energy for our search. Fall out," he says, putting the microphone back.

On the side of the road there was nothing but a dead forest ahead a few dozen feet, but a large clearing of just dead grass and gravel. That's what we parked next to on the street.

"Everyone out," Dominic commanded.

Everyone, including the people in other vehicles, stepped out and started speedwalking into the beginning of a formation. I gathered my collectiveness and my gun and followed the same order.

"At ease!"

This time, his voice wasn't as loud. But it did quiet down… the footsteps were the only thing I heard as the platoon stepped into our parade rest form.

"Before anyone asks, no. We are not setting a camp here. You will be given sleeping bags, and you will find a spot to get some shuteye. The only thing we will be building is a fire. Fall out and gather anything that burns in a pile here," he yelled, pointing to where he stood.

I went back to my car after we fell out and grabbed my bag. Then I started looking for dead branches. In a zombie look-alike world, it wasn't hard. I managed to gather a handful, and I threw it into the growing pile.

A couple of hours went by. Besides us having to do a hundred push-ups because two soldiers started fighting, nothing happened. The fire began to die out, and we were commanded to go to bed at the hit of *10:20 PM*. I didn't want to waste my emergency rations if I wasn't hungry. So, I passed on the cooking of my MRE and went back to my Humvee. Going on the roof of it with my issued sleeping bag. I gave my personal one to Alice so I hope she is using it. Plus, a literal fight broke out between two soldiers because of food. It's not my style to…

"*Private*!" The Colonel yelled.

I practically jumped and broke my neck to see Dominic looking up at me. His arms crossed and his face stern. Like, more stern.

"Unlock your car. We are going to talk inside."

September 9th, 2035

12:30 AM

Capital interstate road field. Texas

What's one thing you say to a soldier to make him panic?

Nevermind. Bad example.

Basically, one of those things is, *"Do you know why I called you in here?"*

"No, sir."

"I would be surprised if you did."

Yeah, he's trying to scare me. How do I know? It's working.

"Where are you from?" he asks. Catching me off guard.

"Abilene. Sir?"

"Are you asking me a question?" he says after hearing my tone.

"Abilene. Sir." I corrected myself.

"Your ethnicity?"

"American and German, sir."

I can see him looking out the windowless front. Thinking.

"Can you speak it?"

"German?"

He looks back at me as an answer.

"A little, sir. I mostly know Zhuyin.

"Chinese?"

"Yes, sir."

"A keen language to learn," he remarks. "You know Monroe is from Hong Kong. Try conversing with him sometime."

"What did you want to talk about, sir?" I quickly ask, aiming to steer away from that subject.

"Why did you defy your mother?"

Was I about to be lectured on how I should be talking to Mom? How does he even know what I did?

"I didn't defy her. I conversed with her," I said briskly.

"Pretty loudly. I heard you on my patrol run."

Yikes. Maybe he was not the only one.

"Ah" was all I could answer with.

"Why did you go against her if she said you're not allowed to go?"

"I need to get someone for both of us. And she understands that."

I wonder how much he actually heard. Because if he was listening closely he would have already known the answer. Though some sergeants like hearing the answer out loud.

It's not like I'm keeping the answer away. I'm just… you know… keeping the answer away.

"Wouldn't you, sir?"

He grunts and mumbles in return. "Like hell, I wouldn't. I would never have that chance. She would have whooped my ass till today.

I fought back a tiny smile. I dont know how I found it amusing, but I guess I lost the fight. Because Dominic sat up and said, "I would make you do push-ups, but we need your full potential tomorrow."

I gulped an answer back.

He puts his big gloved hand on my shoulder and looks at me. Like, at my face.

"You put yourself on the line here, and I promise we will get you what you need. Got it?"

"Yes, Colonel!" is all I answer with, regarding that even if I wanted to say no, I still wouldn't know *how* to.

"Fall out and go to bed."

I open the door and step out, hearing him do the same. I wonder where he will sleep — if he does anyway.

I climb onto the Humvee roof and ball myself up inside the cheap sleeping bag. I tried sleeping, but all my body response was just twists and turns, trying to get comfortable. I would want to count stars to help distract myself, but, you know.

It takes a while but I finally find a barely manageable position to get comfortable in.

6:00 AM

A fantastically horrible sound woke me this morning. Not the sound of an alarm clock or a soldier's voice but the *Pow!* of a gun.

I scramble awake. Raiders? *Sentinels*? Like an idiot, I realized I left my M27 inside the car from my talk. I instead fumble with the thigh holster.

"Everyone up and at em'! Move, move, move!"

I look up to see Dominic with a Desert Eagle pointed in the air. Smoke steaming from the barrel. A totally different man from the one I talked to last night.

I get out of the issued sleeping bag and holster my gun. I roll it up as I jump off the Humvee, stumbling with the bag's unnatural balance, though I make it inside and stuff the sleeping bag in between the front seats of the Humvee.

I step out only to be jumpscared by Dominic, standing six to seven feet away. I swear this guy can *teleport*.

"Good morning, Private! Are you ready to leave?!"

"Yes, sir," I say with a salute.

"What?!" He yells like an old man with hearing issues.

"Yes, sir!" I yell. And he smiles.

"First Sgt! Second Lieutenant!" He yells. Spinning his finger as he walks to the back door of my seat. The driver's seat.

Monroe and Ivys walk by, and while Monroe makes it to his seat and gets inside, I go to mine as he whispers slightly to me, *"Zìshā."*

I know enough Chinese to understand something down the line of *death*.

Finally, all the doors close and I start up the engine. Looking in the rear-view mirror, Sgt. Ivys was as happy as a soldier at war. Dominic is still stern. Okay, just imagine everything Dominic says is stern. Okay?

Though I know he talked to Monroe, which was not a good sign. Though all I could do was move past it. And choose my battles wisely.

It was Dominic again. Grabbed the microphone and stuck his thumb on the gray button. I can hear plastic whine as his meaty grip holds it.

"Radio check echo team!"

This time, it was Ivy's husky scorpio voice. I realized that Dominic handed it to her. So she was the one giving the commands now.

After the unforgivably long radio check, Ivys wasted no more time and started talking.

"Get ready, Roaches! Once we enter Worth, we will not be safe, even under darkness. Sentinels are not the only thing out here. The plan is to get inside the military complex and shoot a giant flare in the sky for civilians inside to spot. We are not leaving for three days or until everyone who wants to be found *is* found. Roll out!"

She hands the microphone back to me and puts it up as all the engines rev up and start driving onto the road again.

6:20 AM

Nine miles down the road, we start to see the tips of the fallen buildings of Worth. Though they were still, in a way, standing.

Some, of course, were down. Some were leaning against one another, and some were rubble. But hey, it's Dallas. John F. Kennedy died here and changed the world as we know it. Maybe even causing this event.

We drove a little longer until the actual aftermath was shown from under its cloak of rubble. Bodies crushed from falling concrete, some burned and downed from the occasional meteor. The only thing showing under them were limbs and skin, skipping the gore. Even passing the rubble, there were fallen people. Not from meteors, either. They were men and women with bloody holes inside their bodies, showing all around their torso and skin.

"Eyes forward," Dominic said, seeing me distracted by the horrid sight. I dont know if he saw on my face that I was about to vomit.

"Two miles until we reach the base."

"What kind of base?" I ask, trying to distract my mind. Desperate.

"An Air Force base. Fort Tokes. It's possibly abandoned, so we can use its supplies and whatever else is available. Possibly vehicles for civilian transport."

More than enough info was provided. My mind was full of not fear. But worry. What if it's *not* abandoned? Who would we meet?

After passing all our obstacles, including a super bumpy road of rubble, we made it to the blown-up fences of the base, Which I'm pretty sure was not blown apart by a meteor.

"Radio a halt. Call out cruiser three to drive inside and tell the rest of the convoy to stay out and stand guard," Ivys commanded.

I did, and after a session of weaving between the vehicles, I pulled inside with another Humvee, parking outside a crumbled yet still-standing small office. Over the top the naval flag is painted on the concrete.

When we got out, we didn't fall in. Instead, it was much more complicated. It was a mission briefing.

"All of us, sir..?" I ask, worried.

Then, he challenges me with a haunting question.

"Why? You comin'?"

After the talk about how we were going to storm inside, I was scared shitless. But after that comeback, I felt... more than challenged. Like, I was *disrespected* for what I was doing.

And that's just not how I roll, you know?

September 9th, 2035

4:50 AM

Plano City,s Texas

"Tsk, what a waste…" Trevor says to himself. He looks inside a dark green tent and keeps repeating "Waste" over and over again until a soldier kitted in black tactical gear carrying an SMG walks up to him. Trembling slightly.

"Ex… Excuse me, beast..?"

"CALL ME SIR!" Trevor yelled, not turning away from the tent.

"Our troops… they dont have any food or water… and-" "What am I?! Your *Mom?!* If you're so hungry, then why dont you eat the food from the jail cell you came from?!" He yells, cutting him off.

 "I'm sorry, sir. Ill-"

"GO MAGGOT! ROUND UP THE SOLDIERS!" Trevor rips from his voice. "We will be raiding the little American base of soldiers helping civilians. Then you can eat the corpses or whatever they have in their pockets! GO! GET OUT OF MY SIGHT!"

The soldier ran off. Yelling loudly, gathering soldiers from all around, possibly to not looking toward the Beast's direction for the rest of the *Month.*

"Your head is mine, Dominic," Trevor says to himself. "I'll make you a wall piece."

He zips up the tent that hid inside the dismembered body of a soldier bleeding out entirely.

He walked out to a fold-out table that had a bunch of armoury tools. He took a Molotov and lit it on fire with a torch that was nearby. Throwing it at the tent laughing maniacally as the blaze reached the dark night sky.

September 9th, 2035

6:30 AM

Fort tokes, worth. Texas

One thing I didn't expect to do this year was to be with a muster of soldiers infiltrating an abandoned military base. Yet here I am, geared up and holding my M27 for dear life. And crossing this off the bucket list I never made.

We stormed inside the main entrance and now we were going down the dark hall. It was only faintly lit by the entrance windows but the light was disappearing quickly as we moved. The light wasn't much anyways, with the blanket of horrible purple candy clouds covering the sun. We moved and stormed every room in the hallway together.

Well, *almost* together. I mostly stayed out of everyone's way. Though my muzzle never lowered.

The soldier, the one they call Ijin, pointed at me and at other soldiers who were with him and gestured with two fingers pointing left. The hallway kept going straight, but there was a spinoff on the left. It was going into complete darkness.

Isn't this the part where most horror movies start?

I follow the two bigger soldiers and Ijin down the dark hall. They started turning on their flashlights that were rather on their shoulders or on their guns, than in their hands. Mine was in my pack so I was relying on their lights. A much muscular-looking soldier, one of the two, pointed at me. He had ripped sleeves of his maroon shirt yet no top gear on his arms or torso aside from a shield-looking armor piece on his shoulder. He points at a door on the left, I walk up to it as he walks up to another. All of us were in front of each door in the hall. And the man with the ripped sleeves stuck up his hand. Flicking up 3 fingers,

Oh God. A storm in. *Alone.*

If I was not sweating enough before, I'm sure I was now. What if someone *was* on the other side? I have no real training for this!

I hear the other soldiers open their doors and I flung open mine, not thinking. Barrel up and ready, it was pitch black inside, so I couldn't see anything. I quickly put down my bag and pulled out the flashlight, still with my gun raised — barely with one hand.

Next time. Keep a flashlight on my person.

It was not a good one, But it worked. Dimming the inside of the room, I saw that it was just a messy server room, aside from one key item. A dead and smashed vending machine was inside. Everything inside was missing, and glass shards were all over the floor, along with wrappers of chips and two full bottles of water. The machine had been fully cleaned out. I grabbed the water but stepped on a glass pane that shattered beneath my boot, causing a loud *crash* that echoed all through the room.

"Weapons live!" I hear one of the men yell.

"No! No! I'm just looting! All clear!" I say, grabbing the water and chugging it. Throwing the other one inside my bag and hoisting it back on my shoulders. Stepping out nice and refurbished as the other two watched me retrace my steps.

"Water?" I offered.

"We're clear. Let's go back," Ijin commanded.

I could only nod in response. Even if I had any word in this, I would have done the same.

We walk down the path into the very dark but dimly lit hall and make it to the other side of the entrance. An exit was provided, and the clearing of the yard was visible though there were bodies outside except not from natural causes or the environment. Bullet holes lingered in all of them. Half were army-kitted soldiers, and the other sum were Sentinels.

I hadn't heard a gunshot since we arrived. We didn't kill these men. Maybe someone else did. Or they did that to themselves… who knows?

"Clear, sir," Monroe says after the Colonel gave a rough inspection of us.

"Good. Search these fallen soldiers for keys to those," he points to the corner of the fence, where two tarped carrier trucks rested next to a small hangar.

As the other men trotted off, I stayed behind

"Something wrong, Private?" Colonel asks.

"Why dont we take the planes?"

I noticed a row of F-15s. They still had a tarp over them, but that's understandable because they were only developed a year ago. It's still surprising when you see two lines of four of them perfectly unscathed, aside from one which half was destroyed by a meteor. Damn, these meteors really hit *everything*.

"Do you know how to fly Private?" he asks.

"No, sir."

"Do you see any planes around in the sky?"

"No, sir."

"Then we dont need them. We need cargo and gas. That's it."

"Yes, sir," I say, moving myself away from him, who just turned his head to bark more commands. There were other soldiers standing together, talking by the exit of the building we came from. I walk and stand by them, all of them bigger and bulkier than me. Yeah, I'm fit, but not *this* fit. It was like each one copied their own body styles from *Rambo*.

"...that's what I was saying. During the morning, I guess they must've killed everyone here."

It was Ijin talking. He was a tall white man, slightly aging but, of course, still battle-ready.

"Hey, Ijin?" I managed to spit out.

"Huh? Oh, what do you want?" he says. It was said slightly with a tone, possibly because I cut him off.

"Who did you say killed all these men?"

"The Beast, I tell ya, Beast."

"Yeah... who exactly is the Beast?"

Everyone's eyes looked at me. Some agape mouths, and some were looking at the floor, while Ijin just sighs. "The devil of man."

7:20 AM

"Trevor Boyten is one of our troops, too. Or I should say, ex-troops."

"He was a rebel, too?" I ask, astounded by the past two minutes of lore that was dumped on me. This guy was a top-class athlete and hand-to-hand combat sergeant. Well classed and mentally insane.

"Like I said, he was the best in our platoon, even getting close with the Colonel," the man with the ripped sleeves said. His name is Mack, by the way. One key thing about him. Very loud.

"He was the best in our platoon, even getting close with the Colonel himself as I said... which now thinking about it… would have been bad news in the long run.”

I sat down on the floor. And surprisingly, everyone else did, too. Like we were telling each other horror stories.

“Until…” Ijin picks up from Mack's place. “One day, we were all called out of the blue to line up at our PT field. That's when we see the Colonel there with Beast. However, he had two swat team specialists next to him. Beast was bound in heavy-duty cuffs. And he had a black leather muzzle on. One that dogs would wear, you know?”

“What did he *do?*” I questioned. Frighteningly curious

“Exactly what you think,” he said solemnly. “Went for the neck, too.”

Why? Why would anyone not only kill with their *teeth* but go for the *neck?*

“From what I heard…” Macs takes over. “The guy that he killed, they were arguing an hour beforehand about allegations. True allegations on whether Beast belongs in a mental facility. So that night, when Surgio tried going to sleep, well, that's when Trevor attacked him. Wildly. I saw the blood myself when we went back to our bunk!” He yells out. Louder than his usual loud voice

But damn, this guy is an animal. No wonder the nickname “Beast” comes to their minds.

“Yeah, but like, if he was an ex-soldier, then why did he kill these soldiers? And civilians, for that matter.”

“When those damn asteroids hit, I guess that's when he broke out. He was in their white box, you know? Where all the koo-koo's go. He rounded up the prisoners inside that did survive and used the police gear or whatever else they could get a hold of to round up his own army he named “the Sentinels.”

“Where were you, kid? In a coma or something?” Ijin says in a humorous tone like he didn't just say what happened *exactly.* Even though most made sense, I still had lots of questions.

“*Fall in!*” Dominic's voice boomed, cutting us all off.

I guess they are gonna have to wait.

Immediately, all the soldiers around me, including me, got up and ran to him into a quick formation. I was one of the last ones on the left of the lines.

“All right, in case any of you couldn't guess, *Yes.* The Sentinels were here, and they took expensive resources. Although, they didn't take everything. Such as this war flare!

He points to a motor launching looking device with propane tanks strapped to it that were all red. They must have pulled it under a tarp because I would have noticed it, mostly because I had never even *seen* one before.

"Essentially the big bad brother of the flare. These are a newer development of a project that wasn't announced to the public. Mostly, the reason is that we never really needed to use them. Yet here it stands. Some of you boot camp roaches know the protocol already. But teach it to a fellow comrade. They can be seen up to fifteen miles away *in broad daylight.*

Is this thing a damn nuke?

"We will shoot it at night, 0800 hours. These clouds are not going to clear up anytime soon, so it would be better to shoot right through them. All of Worth and Dallas will be able to see it because luck has it enough. We are right on the outskirts of Worth that reaches Dallas. Does everyone understand?!

"Yes, sir!" We all chanted.

"Fall out. Ijin, Monroe, help me set this flare down."

September 9th, 2035

8:55 PM

Fort Tokes, Worth, Texas

We stayed at the base for the whole day. I mostly ate and cleaned up. I was dragging a body to another grave we dug for them. We had lights of our headlights and a large fire pit filling up the base. Everyone moved inside with us, and Ivys was eating a MRE at the fire with the other soldiers, silently. Dominic was standing by, crossed arms and serious as he watched us finish up burying everyone.

God, I was so tired. Just a side note for you all: a dead soldier weighs a *lot*.

I had my bag by the exit side of the building we raided. So it was a little easier. But still very awkward. It didn't help to know the fact that these soldiers were *dead.*

I dusted myself off and went to my bag. Grabbing another MRE. I only had one today but understand that they fill you up so quickly. I headed over to the firepit made from the tarp that covered the flare and put mine next to it along with other MREs. The owners were all sitting around, waiting for their meal to cook. They watched but didn't say anything. They were all too busy focusing on theirs to focus on mine until Dominic stood up, causing heads to turn.

"Alright, men. Cover your eyes and ears. If any of you miss the sun, trust me. You're about to feel it."

What he actually meant to say and to put it in a "Straight to the point" terms, it was time to fire the flare.

"Firing in ten seconds. Cover up."

I covered myself like I was covering up for a tornado. Reminded me of better. And darker times, aka when I was in the closet… before all this — without worries or cares. At least not as many as I have now.

"Three,"

"Two,"

"One."

I heard a loud *Poomph!,* followed by a fireworks-like noise. The screech it left was deafening.

But I also *felt* it.

You know that feeling you get after opening a burning oven on your face? That feeling was just what I was feeling. I was covering my front half, but I definitely felt the feeling on my back *heavily*. After about a minute, my back cooled down, so I went to look up until another soldier forced my head down.

"Not yet."

I managed to catch a glimpse of the beam of light that shot out like a beacon. I would be surprised if it didn't leave a hole in the clouds.

Another minute passed until one voice yelled, "*Clear!*"

I stuck my head up to realize it was Monroe who pushed my head down. To my surprise, for sure.

"Thanks…"

"Dont do it again," is all he responded with, while he laid back down.

"Batteries off!" Ijin yelled.

I got up along with a couple of soldiers and, walked over to my ugly-tagged Humvee and went inside. I turned off the headlights and grabbed the sleeping bag that was shoved in between my seats. I then moved back to the slow fire and grabbed my MRE which was now charred a little.

Still ate it. Chicken Alfredo is really good.

11:23 PM

I was falling asleep until a loud gunshot awoke me. Though, Unfamiliar. Not Dominic's handgun, Quieter. More of a sleek sound rather than an echoing *Bang*.

Nevertheless, everyone was up — guns drawn to the new sound.

What was I doing?

Looking for my M27 frantically, I settled with giving up and pulling out my handgun. I looked around and pointed it at where all the barrels were pointing.

And at the broken gate of the fence, we made to move all the bigger trucks inside stood the figures of a man and a woman 20 feet away.

The woman was a short yet beautiful-looking dark woman. There was barely a scar on her figure. Lean yet with curvature. However, she did cover herself with winter gear as if she were a homeless person on the streets of New York. They were navy blue and brown, matching each other.

While the man was the polar opposite, he was a tall, fully golden blonde. One you would see in summer commercials. He wore a tank top and cargo pants that were cargo gray with a digital print on them. His tank top was trying to hide a large, sinewy, sunburnt figure, though he stood around 5'10 — only a few inches taller than the woman.

The woman carried a wooden sniper rifle, pointed in the air. The muzzle smoking.

11:24

"What ought to be the meaning of this?!" Dominic yelled as if someone had shot a gun to wake him up. "Who was on watch?!" he says, crossing his arms while walking through the troops who were all armed and ready. Lowering their barrels but still at the ready.

"Our sincerest apologies." The woman said loud enough to be heard by all of us. She lowered her rifle and strapped it to her shoulder. As she did, she spoke in a contralto voice. Matured and understanding. "We saw your flare, so we thought you needed help."

"We shot the flare to bring anyone who needed help to *us,*" Dominic said, cold and unfazed.

"And who would "*Us*" be?" she pushed, tilting her head slightly up.

"We are the rebels. A PMC company by the government of the United States who was made to train troops for a fallen world. Also known as us.

"Trained? You knew this was going to happen?"

"Negative, ma'am. We have been in service for eight years."

"That means someone in that damn oval office *did*." She sighs. "Well, I'm Zu. And my buddy here is named Troy. We came with the intention of saving you."

The silence was all that followed.

"Are you Sentinels?" Troy said, breaking the tension of silence.

"Negative," Half the crowd answered.

"We have a base in Dallas. However, it's fortified so we will have to go back with you.

"How can *we* trust *you?*" Dominic asked with a tone of distrust. "You could be leading us to the Sentinels."

The woman, Zu, pulled a blue flare gun from her jacket. And held it out like it was an impressive art piece.

"When I shoot this flare at a specific time, you will see another in the air north of here. This will indicate that you guys are safe and you are not a threat. All we ask is to camp with you."

"And how does that prove your trust?" Dominic asks.

"Easy. If I take you tomorrow, you can see the camp for yourself. You can put us under watch if you want. We will take you tomorrow morning."

The Colonel took a short time to consider it. It's not like there were any better ideas. Plus, if they were Sentinels, they would have attacked already — flanking our sides or behind. I'm guessing that's why Ivys was looking the other way. Inspecting.

"Alright. Though We dont have a cot for you both."

"We brought our own."

Then she points the flare up and fires it. Moments later, we could see a similar blue flash in the center-ish of Dallas.

So little words were exchanged, but 30 minutes later, we had a plan. A few of our troops — if these people were telling the truth— were going to escort the civilians to the school in Ranger. Simple enough, really, though, we still had our own mission to go to Houston and activate the beacon alarm.

Although I would be selfish if I said everything went *all* smoothly.

I left my bag and AR leaning on the wall of the naval office — and as punishment, I had to do a hundred push-ups and be one of the lookouts for the rest of the night.

September 10th, 2035

6:00 am

Fort Tokes, Worth. Texas

Today started with another gunshot from the colonel. And curses from the soldiers. No one else came that night, I should know because I was awake. Insomnia is a bitch, but it is sometimes helpful. Such as tonight.

Didn't mean that I wasn't tired.

"Fall in!"

I grab my M27, get up from my Humvee where I was sitting, and run into the formation where everyone was. Except for the commanders and the two… who were they again? Zu and Troy. Right.

Both of them were looking equally as upset as the rest of the soldiers, though they were packing up as soon as they got up. Ivys walked up to them and started talking with them. They were out of range for me to pick up anything, though.

"All right men, as you guys figured out last night, we will be following these two soldiers to this… camp. So, everyone keeps to code fifteen." Dominic says

Code fifteen?

"Everyone understands?"

"Yes sir!" everyone chanted. Including me, even though I didn't have a clue.

"First Sgt!" Dominic yelled as Ivys walked over with the two.

"Take charge of the company," he commands. Stepping to the side as Ivys takes his spot with the two on her side.

"Now then, let's get started," Ivys says, putting her hands behind her back. These two will be taking the carrier truck. We couldn't find the keys to the second one, so let's hope everyone fits in all three."

The newcomers stayed quiet. Now over their groggy temper from the awakening, they silently moved away and went to the truck. Ivys pointed at two soldiers and gestured for them to follow the two. Guess they are riding in the back.

"Roll out!" First, Sgt yelled. Turning around and moving away from us as we all made it to our vehicles.

These commands hit our heads now like a Malace, and it hit well because everyone was pulling out of the base in less than ten minutes.

"Light!"

I practically jumped out of my seat. It's not like he is sitting right *behind* me.

"Pass over the microphone"

I quickly grab it and hand it to him. I didn't even feel his hand, yet I sensed his tight grip around the square microphone as if the crying plastic was going to crush any second.

"Radio check, echo team."

After a radio check, including our new cargo three, Ivys grabbed the radio from Dominic.

"Cargo three, lead the way.!"

6:42 am

You know road trips? Like the boring part? Where you're just driving straight on the road in complete silence? That's all I felt when driving to Dallas.

The only minor difference is that 90% of the world is dead. The other 10% is destroyed. Physically, in some. Mentally, for most.

And I was both.

Oh, and the windshield was smashed open, so it was kinda loud.

Besides that, it gave a classic road trip vibe.

Though I was getting the hang of the Humvee, steering almost felt natural. And I started getting the ache to drive it faster. As fast as it could go, though with the traffic of units forward and behind, it wasn't possible.

"Lock the windows," Ivys said from behind.

"But we don't even have a *windshield*," I say in a slightly complained tone.

She glared at me, and I felt her gaze bore into me. I instantly fumble with the window lock, and when I finally press it. The metal rod bolts close and pinches my finger slightly.

"Fuck!" I hissed at myself.

Monroe silently snickers at me. as I embarrass the hell out of myself.

Two more minutes down the ruined road, we begin to see a barricade, not of rubble or trash, but instead, well-equipped Hummers and Growlers, two of each making a line of four. One of the Humvees even had a Grenade launcher in the back.

Suddenly, the Humvee in front of us came to a halt. The line was frozen,

"What's happening?"

"Zu is leading. So, she must be getting their attention."

As if on cue. Whatever she did worked. The trucks moved out of the way, and about half a dozen soldiers dressed in Air Force ground unit gear walked out. They all had blue tape wrapped around their biceps, and they came out guiding each vehicle inside.

You know, when we drove in, I was kinda expecting at least *some* kind of organization. Instead, there were just *tents,* all sprouting from random places in the cracked sidewalk, some reaching the asphalt along with a few barrels of fire.

Just the physical sight of this is enough to break a pure soul. But the images shown on the faces were more heartbreaking; scarred beauty was unnatural when it clung to the human skin. Families and friends clung to each other. Some cut, some bruised, some lost more. Amputation was an uncommon thing in the world usually, but now it seemed more rare to find someone fully intact. Even the soldier dressed mostly in tan gear with a similar blue-wrapped bicep was missing two of his bottom fingers on his left hand, though he still carried an M-17 Scar.

This dystopia reminded me of our home camp.

The soldier walked up to the window after guiding us a few more feet and tapped on it with the back of his hand.

I fumbled with unlocking the window until Dominic just opened the door and stepped out. Slamming it behind him.

After watching them for a minute, Dominic turned and knocked on my window. Luckily, I managed to figure out the stupid window beforehand, and I rolled it down.

"Park. We will be loading down where you are. Take your keys and keep them safe. We still don't know where we are.

I nodded for an answer before tightening my bag onto my back again and rolling up the window.

Oh yeah, that will help, especially with *no windshield.*

"Don't pinch yourself," Monroe snickers as I lock the window again

"Easy for you." I sneer

I open the door and step out. Slamming the door like Dominic. Only a little harder to shut out Monroe's voice. Please take the hint and leave me alone.

"Fall in!" Dominic boomed.

I ran to the troop who were already in a starting formation line on the left side of the convoy. I fell in somewhere between the front troops. Of course, ahead was Ivys, though I didn't see Dominic anywhere.

Never mind, he was walking up with Zu. with ten soldiers following behind them. All kitted and ready for anything.

"Formation!" Zu yells out as she walks. The soldiers accompanying her were now running to a separate line next to us in another squad. Two of the ten ran off to the tents,

"Take charge, sir," Ivys said to Dominic. Saluting. As Dominic saluted back. Taking her ground as she moved away from it.

A few minutes passed, and the soldiers were running in from Zu's army, slowly filling out five rows of soldiers. A small cross count later told me that they had seventy soldiers. Honestly, I thought this army was more petite, Way smaller. They managed to make our army seem *dwarfed* with comparison, though they were substandardly strong. A couple had SMGs, while others had ARs or Shotguns that were rusty or brown. They must have gotten the newer stuff from the base while the low-quality weapons from the prisons or wherever else. They were mostly geared well, though all dressed differently, their only resemblance that they were in the same army was the patch of blue tape on each of their arms. Distant chatter emerged between ourselves and them as we stood.

"At ease!" Dominic boomed.

After everyone got quiet, he began talking, loud and proud.

"To the New Frontier Army who don't know me, my name is Colonel Dominic. Commander of the First Independent Ruins Army. The United States government made a PMC to have troops ready for times like these, the Dark days of today. The man you see standing beside you is our fourth battalion of roaches. We are here to aid you all while we bring you to our base of operations of another thousand or so people we have already, where you will get rations and medical supplies if needed or wanted.

Among the mixed soldiers, one yelled out, *"Looks to me that you're the one in need of help!"*

Then the whole army started shouting out a couple of *"Yeahs!"* And voicing out tones of agreement, causing the frontier soldiers to grow slowly louder.

"Silence!!!" Zu roars.

How the hell do they *do* that?

This time, Ivys spoke.

"We understand and see your points and worries. But our army is well-trained and equipped for this kind of combat. You all, if you're from the base that is, are trained in aerial combat. This is far more extreme than a dog fight. These men of Beast's army will kill you slowly. And they have the manoeuvrability that you have. It's not about the better equipment now but the better soldier.

"Do you know how hard it is to fight a dogfight?! How can we even trust you?! You could be allied with the Sentinels!" A different voice yelled along with the hums of concerned soldiers.

"The only way we will be able to actually prove it is if you join the Rebels and follow with us. If you can do that, we will take you to the base." Ivys says.

Shouts broke out, angry shouts. All of the Army was yelling at him. It was hard to keep my head forward and not look at the rioting crowd next to us.

"We don't even know you!"

"You could be sentinels *yourselves!*"

"This could be a trap!"

All kinds of shouts steamed through them, all shooting at us to answer.

Finally, Zu stepped in.

"SILENCE!!!"

Her shout, in which somehow is louder than before, causes the angry army to fall quiet

Zu then walked up to the middle line of where both of our platoons were and shouted loud enough for us *all* to hear.

"We just met this army and they want to recruit us. Ha! At least you won't have to prove if you all are *American*. She laughs. "See, we don't do *"Politics"* or *"Elections."* We do survival of the fittest. So I'll tell you what, if you, and your army beats *my challenge* with *my* army, I'll allow you to command my troop till they dance on their heads!" she says. Sneering and eager.

"We accept." Dominic said, almost instantly.

If he has so much confidence in this troop, then I should too right?

I didn't know why I didn't.

Honestly, these soldiers that we have were *soldiers*. Like, actual *military*. And they were Spec-Ops elites. A private company. Quick with command and quicker with action. If anyone could call themselves the "real army" here, it's the rebels.

"Good." Zu says. "Now for the challenge I need to get three soldiers, then you need to pick one of mine."

"Pick anyone you like." Ivys says, stepping in and gesturing her hand to us.

"Hm," she thinks for a minute before. "You." She points to Ijin as he walks out of line to the front.

"You," she says quickly after, pointing at a small, dark-skinned lady. Causing her to go next to Ijin.

And hey. Wild card for anyone who guesses who she picked *last*.

Take a wild guess.

September 10th, 2035

4:00 pm

Frontier Camp, worth. Texas

So, I failed. Not only myself but my platoon.

7:20 am

In short, here is what happened.

Zu picked me. (Wow, shocker)

Ivys definitely didn't pick the smaller guys of the other team, that's for sure. But anyway, before we get to mine, let's get to the first two.

The first one was a marksmanship challenge. And Ijin was in charge of that one. Getting us an easy one out of three victory.

We messed up on the second one. "Rose," the soldier that was picked up with Ijin and me, couldn't start a fire as fast as the muscular white man that she went up against. So, so we lost that one...

Making *me* the tiebreaker, to a grappling match.

And it was very embarrassing. I got thrown on my back the first time, though, in the second round, I managed to crawl over him and pin my whole-body weight onto the back of his neck. Making the final match very unexpected; in other words, I won.

Everything was *on me.*

Ten years of training did not give me any hope.

When the gunshot went off for the last match, he charged me. Mind you, he is a 195–200-pound man. Pure muscle and weight. So, him grabbing my shoulder and flipping me over him was lightwork. However, when he did, I managed to land on my bottom instead of my back, causing him to jump on me. I slipped away and wrapped one arm around his neck, crawling on top of his back. But he slammed his palm into my elbow. Loosening my grip and giving him the advantage to grab me and slam me down with his knee pinning me.

The difference between us is that I was trained to fight.

He was trained to kill.

And so, I lost, along with our army.

Meaning?

We were (for now) under the command of "The New Frontier Battalion Army."

It wasn't all my fault. *Ish.*

Actually, no. It was all my fault.

As I fell back into my spot in formation, I could feel the laser eyes burrowing into me. I wanted to say something, but I *couldn't.* They were all right to do so.

"As per our agreement. Your army will serve under the protection of *The New Frontier!*" Zu yells. Loud enough for our commanders to hear. "And as such, we will not be going back."

"As for us. We will be doing the same." Dominic said after her. "We will try our soul hardest to serve and protect these new allies of ours and serve to share for them our security if they do choose to change their mind."

"While you do. We will be evaluating your actions, then we *may* send civilians into your hands."

Our army side stayed quiet.

"You heard her!" Dominic yelled. "Pack down wherever the Frontier Army assigns you. The troop will guide you. Fall *out!*"

"Nice going, *momma's boy,*" Monroe says. Shoving me to the ground as I walk past him.

So, I ignored it and speed-walked to my Humvee.

4:23 pm

So where was I? Just huddled in my Humvee, parked in our assigned spot, fighting tears and anger.

Damn. I just want to *break* something.

I grab my bag from my passenger seat when throw it at the window, and rummage through it. Pulling out a flare, then a compass, then an MRE…

Growl…

When was the last time I had a meal? Since the night of the fort? Practically a whole day. I'm told food calms a crowd. It should work for me, I'm hungry anyways.

I step out of the truck with the MRE and walk up to a burning oil barrel. No, it's not oil that's burning, I don't actually think that's good for food. Though it was cut open to put wood inside, the top was mesh. And it was already lit and ready for use. I throw on the MRE and the chicken wire mesh catches it. There were men around, standing, sitting, talking. Yet when I came out, I got glasses and whispers shot out in between them. Like it was nine *hours* ago, move on.

I moved to the sign that was where we were. Yeah, our assigned spot was a parking lot. But we all could fit, and we had some shade and comfort. Concrete walls of fallen buildings surrounded all of Dallas like a maze. The sign I leaned on only said Arlington Community Parking.

Wait.

Arlington! Holy shit! I'm like- Here?! I neglected to even remember the only reason why I'm here! I shoot myself away from the sign and dash to look for Ivys. definitely not Dominic. Especially from my embarrassment earlier? Yeah, I'll be avoiding him for the next *century*.

We were not that far from the makeshift camp they had. Originally at first glance it looked like a street that was loosely blocked off, disorganized would be the right word. But in fact, it was organized as organized as a robust base could be. There was the roadside. which was where all the civilians had their tents. A soup kitchen-like area. Four Humvees on one street end while the other had three.

Simple enough.

It was still so surprising —like at least 100k people lived here, in the city alone. And out of those hundred thousand, I would only most likely see not even 5% of that number.

Dominic most likely is with Zu somewhere. Though, Ivys is a different case. I went to the concrete courtyard in front of a ruined building, the place where the grappling match happened. There were a few Frontiersmen sitting by a broken-down statue. It may have stood representing someone strong, or important, smart even. But now, it was just the legs of a business-class gentleman that stood alone. The body was seemingly chopped in half and smashed somewhere in the scattered remains of other stones.

The floor had random white parliaments on it. Originally I thought it was snow or something, but with the air and sky like how it is, I realized that it was *ash*. Fallen from the burnt-up sky combined with the ground. It didn't affect anyone else seemingly, so I acted as if it didn't affect me.

"Private Light! What are you doing here?!"

I spun around startled. Not thinking as I see shockingly, Colonel Dominic was standing behind me about four feet away. Of course, as intimidating as ever.

"Ah! Hello Sir! I was actually looking for you!" I blurt out. Quickly

"Private, I don't want apologies I want *improvement*," he says.

Getting the wrong idea…

"What? Oh no sir, I wasn't going to apologize…" I paused after I saw his face while releasing what dumb shit I just spat. "Wait! what I mean to say is I was gonna… but- I… I was- um… well-"

"What do you need, *Will?*" he asks rigidly. Rubbing his face. I must have already annoyed him because he used my first name.

Over that fact, though-

"The reason I am here," I explained, pointing down the street that was blocked off by three vehicles. There were two Humvees and a growler.

"Your broth- I mean… that person is out there?" He questions. Surprised.

Yeah, he heard everything from that night.

"Yeah, I assume so… There is a hospital over there named UT South Western Medical. It's only a few miles out.

His face tenses. He is thinking.

Even if he wasn't going to let me go, I was going to go anyway. On my own if I had to.

"I will allow it. But not today. Seek out the sunrise tomorrow."

He must have read my mind.

"Sir, if I may, I am very much prepared to go now. I am already several *days* late. Anything could've happened, and anything still *can* happen. He can't-"

"Understand it's well past afternoon and we have no idea if this building of yours is still even *standing*. Plus, with what happened earlier, I'll have to notify commander Zu." he cuts in.

I cringe from the memory, but damn. He was right. About everything, too.

Even the building being…

No. That hospital has to be standing. Hospitals are really fortified. This one is small. It has to be somewhat decent.

It's still standing.

I have hope.

"Yes sir." I sigh, faking a defeat.

"Good. get back to camp," he says, walking off to find his new commander. I guess ours, too…

And for the record, yeah, no. I'm not going to listen

What? My brother is out there. In *that.* I have no other choice.

I gather my wits and begin my path back to camp. Forming a plan slowly.

Making a plan is easy

Putting it into action was a different story.

September 10th, 2035

6:00 pm

Fort tokes, worth. Texas

Seems like these days are just flying by. It's been almost three weeks since the meteors fell.

Yet it only was a few days of feeling for me. And it is still hellish.

All I can't complain about is that I have plenty of things to do, Physically and mentally.

Right now, the mental side of me is full. It is steamed with plans. Plans of escaping.

Actually, sneaking out is a better way of putting it.

Scared? Yes.

Determined? I think.

At least, I better be. Or else this plan will never work.

Back at the camp, everyone was getting ready to clock out for the night. And almost everyone set up to sleep outside. I wondered why, though.

For fresh air?

Whatever reason it was, it made things harder. Actually, it's going to be impossible to drive out in my Humvee; too many people in the way, *and* it's way too loud for anyone to just *not* notice. Guess I'm ditching the car.

Right. Game plan.

So, I'll sneak out on foot. I'll leave my keys just in case the car is needed, plus they won't need to search for me. If I fall behind, I'll be left behind. And I'm perfectly okay with that. I can't be a liability for my troop, not again.

Getting in was a challenge, but leaving was a lot easier. About an hour later after I prep, I sneak off and away to the left end of the street. Two Humvees and a Growler was what greeted me when I walked up, blocking the path. Inside the Growler were two soldiers geared in digital camo; theirs was tan, and they both had M16S'; the one in the driver's seat looked over to me and stood up, peeking his head over the railing of the unshelled jeep.

"Are you goin' out there?" he says. He was a dark man. In fact, he was the man who guided us inside when we first arrived. I saw his left hand, and that told me so.

"Yeah, just for a walk. I need to check something, "I say, resting my M27 on my shoulder. The muzzle pointing up.

The guy bends down back inside the Growler and pulls something black from under his seat, and he throws it to me.

"When you get back, shoot this red flare in the air," he says, throwing me a red flare cartridge. "And here-" he says, throwing me a green cartridge. "Shoot that one wherever you are if it gets too dark, the hour of eleven. it lets us know you're staying out and you're not dead. Though when you do, make sure to walk at least 100 paces away from wherever you shot it. Someone else can see it as well as we can.

Finally, he throws a blue cartridge flare. It hits my head while I'm inspecting the black spray-painted flare gun. I bend down and pick it up.

"The blue one is for when you are in danger. Now, keep note that we won't be able to save you." He says, sitting back down and starting the Growler, slowly driving out of the way as he talks. "When you do shoot it, it tells us that you are rather in combat or are finished. Therefore, it will tell us to ignore any other flare you or whoever else shoots one up. I'm sorry but we need you to keep us safe. If you do end up overcoming whatever challenge you face and make it back, I'll let you back in.

That's comforting, so I just silently nodded in response. What's not comforting is the fact that I won't be able to get help. I'm on my own. However, I was taught in 3rd grade that actions have consequences. And these are mine.

"Alright, thank you"

"Go on then," he says, swinging his arm

I walk past the opening where he just put the Growler back at rest. Fortifying the final brick of the wall of jeeps.

Yeah, that's what this is.

That wall. Now, a fortress ready to keep anything and everything out. All from the dead world on the outside of it. I can't help but stare at the sky, which gets slightly darker every minute. The rubble above and the crippled concrete below it. With the smashed cars that surrounded these once-great buildings.

I focus my mind on the ruined path ahead, head high as I walk.

8:12

I can't see anything useful. It had gotten way too dark, way too quickly. Usually, even the best observers would know exactly where they were up when they had the night sky. But the starless night, along with the shrouded moon, would be the main reason why anyone could get lost. It was close to pitch black out. My only sense of direction was a compass. (sadly, a broken one because I tripped over some rubble twenty minutes prior.)

I was never afraid of the dark, but at the moment, if you saw me, you would think I was. A trembling kid, heavily overgeared with a gun, pushing half my size. No, I was not trembling from the very cold night. I was scared.

Hearing only so little yet so much about this Sentinel army, I was scared straight. They could be out here now. And worse, they could be tracking me, watching my body as if it was a huge bullseye for a shooting range. An easy target.

I surround myself with other thoughts, continuing my path forward, happy thoughts. Of my mother, father, brother...

Really, these were all I had. I don't have anything else I really care about, lonely to say I don't have friends, just all I care about is my parents, Jack, and the gear on my back. My dad's gear, that's still being used to this day, is as worn as Dad left it from his time in the military.

Oh right, I also have the necklace from Alice.

I take it from under my shirt and try to inspect it. It was a golden bronze double-capped vial of straight soil. Both sides were rounded and tarnished. It resembled a clear vial of a small pill. And a white seed could be seen inside the brown dirt. And it wasn't tightly packed, whatever was inside could have some growing room, though it's the size of about 2 inches. I don't think anything can actually *sprout* inside, though I'll find a spot to plant it sometime, I guess this represents the new world in someone's eyes.

What's funny is that it was made a while ago. I would have assumed Alice made this recently because of the situation we were in, but both caps were sealed shut, something that has to be custom fitted by a jeweler to fit as tight as it was. This was an older necklace, why would she have it? She was never really into farming or agriculture. Really, the only thing she liked a lot that grew from the ground was flowers. I'm guessing this is a flower seed of some sort.

I shove it back down my shirt while continuing my dark path. It confused me more than it weirded me on how old the necklace seemed, though I'm sure my mother gave it to me as a sign of home and hope. To replant it when the world is ready.

Keep hope. Keep-

Woah.

I try looking back up in front of me, but I *can't*. A tintinnabulation screech filled my ears for a brief moment. I tried lifting my head and yet I couldn't. It felt as if a stack of bricks were stacked on top of my skull

What the *hell* was happening?

Did I get darted?

They must have hit my neck. The fluid could be spreading in my body now.

Panicking, I grabbed the flare from my pocket, dropping the flares on the floor. I get on my knees to pick one up. I don't know which flare is which. The colours were scattered and mixed in my head. Plus, it was *dark*.

I grabbed one that was… green or blue… I couldn't tell. My mind was blacking out with the sky.

The keening sound got louder.

I grab the greenish-blue one and begin to try to get on my feet. But I couldn't. The feeling reminded me of being tied up.

Never been tied up, but you get the point.

So, I give up on getting up. I load the flare in and slam the gun shut.

The ring fills my ears, very much hurting.

I pointed the flare gun up, and cocked the hammer back.

Click

*Poomph! *

Suddenly, the ringing stopped. And I started dropping; I blacked out before I could recognize the colour of the flare.

September 11th, 2035

4:00 am

Plant City. Texas

"ANNOYING! ANNOYING! SO, SO ANNOYING!!! I will KILL THEM *ALL*!"

Trevor stood on the building roof with binoculars in his hands, pointing at the damned makeshift base twelve or so miles away.

The same soldier that Trevor recognized was with him, the one who annoyingly complained about food a few days ago, with better gear and a less feared look on his face.

"All scouts and units count only one man leaving."

"That's it?! Why are they staying?! Tch!" Trevor yells, slamming his binoculars on the floor, the glass lenses shattering and the stainless-steel bending upon impact.

"You!" Trevor pointed his big finger to the man who was with him. "What's your name again?!"

"Bush, sir."

"Well *Bitch*, I'm ordering an assault! Today! An hour! *GO*!" Trevor barked.

"Sir! That's impossible!"

"ARE YOU DEFYING ME BUSH?!"

"N- no sir!" Bush stammered, quivering with his words. "It's just-"

"I don't have time to argue with you!" Beast yells. Picking up the bent binoculars from the ground and chucking it at Bush's head. It hit with a loud *ting* from his helmet, stumbling him back as blood slowly leaked from the right side of his head.

"We don't have a plan sir." Bush said, quieter as he stood up straight again "it will take at least a half hour to even notify all the army!"

For the rarest moment known, the beast paused. He looked at Bush as if he was a gazelle, while Trevor was a lion who was not really hungry but *bored*.

"Three hours." He finally snapped. *"GO!"*

"Yes sir! Thank you, sir!" Bush then ran to the side of the building and climbed down the rack ladders."

"IS IT READY!?" Beast yelled suddenly.

Silently, a nurse came up the ladder. She carried with one hand off the ladder a rebar handle. A rod about a foot long. When she managed to climb up the ladder, the rod end was welded to a piece of iron cut into the shape of a war axe, and it was glowing hot, an orange tint covered the whole axe as it faded to red, and finally, the rest of the metallic rod it was welded to remained the same silvery dark colour.

"Finally."

Beast walked away from the roof ledge and walked up to the nurse. While the nurse was a five-foot small woman in her early twenties, Trevor was monstrous. 6'7, 250 pounds of pure muscle. His chest was big, but his shoulders were bigger. Arms as thick as textbooks, and his neck was almost as thick as his head. He wore a black paracord necklace that connected to a stainless-steel muzzle that was crushed to an almost unrecognizable state.

As for his clothing, he barely had any on. He was shirtless. He showed tribal tattoos that surrounded his abs and back with varying colours of green, red, and black. With only camo cargo pants and a black bandana. Tying his wild, long, brown hair back. His face is stubble, and his eyes a dark purple, full of fury.

Trevor grabs the cold end of the rod and picks up the branding tool

He stares at the glowing end for a couple of seconds before he turns it over and shoves the brand into the front of his shoulder.

"Arghhh!" the beast silently groaned. Trying to keep back his yell, shoving the brand down with more force as his hand, and whole body starts to sweat. Trevor holds it there for about twenty seconds before removing the sizzling metal from his burnt skin. It glowed, but not as brightly as before. It was a solid dark red rather than a bright molten orange. The axe design was stamped and planted on his shoulder. Clean.

"Gah-" Trevor sighed. Still holding the rod. "Alcohol! Where is the alcohol?!"

The nurse jumped at his sudden tone, answering trembling. "I'm sorry, sir! We didn't have any, but I'm sure I can find some!"

"YOU WANTED ME TO DIE FROM INFECTION HUH?!" Beast ripped. Grabbing the dark-haired woman by her collar and pulling her close to his face. She screamed and tried resisting the big hairy hand, but hopeless as it was, she trembled after failing.

"Please... Please sir..."

"Oh, don't worry, I ain't gonna gonna kill you." Trevor then grabs her jaw and lifts up the scorching rod.

"Maybe looking at your lovely face every morning will remind you to be prepared. Trevor said, smiling before shoving the stamp into the nurse's face.

Apollo Diaz

September 11th, 2035

6:00 am

Somewhere in Dallas. Texas

I open my eyes to the purple sky and the gunpowder air.

Familiar, right?

So, I sit up and look around.

The cracked concrete path, the fallen buildings, the destroyed street covered in ash. I'm still where I was when I got knocked out. Though it was brighter, I guess the sunrise should be showing. Keyword, *should.*

Did I fall *asleep?!"*

I feel the back of my neck, then my arms and legs,

No dart. Not even an open cut.

I have not been sleeping… like at all. All the way from the night Dominic told me to keep a lookout, I haven't slept, in what lookout, I haven't slept in what, like — two days? Maybe more?

I get up and dust off the ash that rested on me. I see the black flare gun, so I pick it up while picking up two flares.

One red.

One blue.

Good and bad news I guess. Bad because if I was in danger then-

Oh wait…

Maybe all good

But I don't think I shot it at 11:00. Will that mean anything?

Possibly, I'll find out when I get back.

I picked up my bag, and *damn,* I possibly packed too much crap.

100

Growlll…

Never mind.

But I can't eat now. I don't even know where I am.

I look around and I spot a street sign. It wasn't helpful much. The sign read "Far-" and the rest was cut off. A giant piece of concrete crushed the rest.

Unliftable.

I'm not even gonna try.

But then I get this… scent.

One that was weirdly familiar. Watery scent. Sweet and thin… Rain?

Wait no… *flowers.*

I grip the concrete as I climb on top of it, allowing me to see over some rubble and a little farther out.

Finally, coming into view, a hospital. With an aura of flowers growing all around it.

I stood in front of the hospital.

The hospital

The one I chased for the past few days. The sidewalk to the doors were the only things that didn't have patches of flowers growing on them. It took me ten minutes to make it to the entrance. I study the view. The watery, sweet scent in the air — all the little things I missed so much.

"Isn't it pretty?" a Scorpio voice says from behind me.

September 11th, 2035

6:18 am

Arlington Medical Research Facility. Texas

I spun around, grabbing my M27 and embarrassingly, tripping over myself. Instincts are good. Clearly, I'm just a dumbass.

"At ease, Private, it's just me."

I look up to see Ivys standing around eight feet away, with a kindred look on her face.

"How did you get here?! How did you even find me?! Am I in trouble..?"

"At ease!"

My out-loud questions stopped, but my mental questions continued as I picked myself up and dusted off my bottom.

"After seeing the flare that was shot last night at a completely *random* time, I called in the troops. Guess who was not there?"

I sighed in response.

"So I figured you would need a ride back."

Huh?

"You're not going to yell at me?"

"Colonel told me why you left, so."

"He knew I left?"

Ivys just nodded in response. "I brought your car." She says, sticking her thumb to the Humvee, but it was different. And Ivys read my face again.

"Yeah, I added a windshield.

"That's not a windshield..." I say, staring.

"It's a makeshift one."

Very makeshift. A steel door with two small eye holes' cuts into it was welded to the Humvee windshield placement. The size of the slits was about eight inches in height and 2 feet in length on each side.

"Whatever works then," I say gloomily.

"These are the first plants I have seen since the event started-"

"I'm ignoring them, I have bigger goals," I say out loud. But in my mind, I was baffled. *How were they even growing?* Hopefully, they don't have some radioactive end-of-the-world properties.

"Good focus," she answered approvingly. "I'll guard the entrance."

"Thank you." I responded. "But with just a handgun?" I gestured towards the holster on her belt. "Do you want to hold onto my AR?"

"Keep it, I work best with single shots, like snipers. Leave your bag. Stay light. I will watch it."

"Thank you, ma'am," I say, walking up to her and taking off my bag. When I hand her my bag, she takes it from my hand. And grabs my forearm.

"Check your weapon. Be safe."

Did she just *worry* for me?

"Yes, ma'am." I salute her and raise my AR. With, for the first time, confidence on my face.

I turn and head through the overpass of the emergency room, going inside.

6:25 am

In situations like this, the hospital is pretty creepy. I don't know what's a better thought. Believing if I'm alone or if I'm not.

Luckily, Ivys is here.

I move past the lobby and into the main hall. There was a map made of acrylic or plastic on the wall. It was faded, and cracking appeared in the corners of it. But it was still readable and small too.

Sorry, Not the map, but the hospital. In general, it only covered about an acre or two.

Is this *really* a hospital for a *city?*

The floor was littered with papers. And I mean *everywhere.* Scattered medical documents, examinations, diagnostics, but I was more stunned by the verdant flora that surrounded the floors and walls. All that filled my head was, "How was I ever going to find my brother's room in *this?*"

Obviously, by looking first.

The map showed that the biggest room aside from the cafeteria and the MRI room was not the lobby, but instead, a room seemingly in the middle of the facility, named "Singularity intensive patient unit." It was at the end of the hall.

I guess that's where he would be.

I walk down the hallway, slowly. My gun still slightly raised. The flowers of all colors getting more, and more vibrant. The doors on both sides all open. I noticed a massive influx of flowers in one room. So, I looked inside.

Big mistake.

I see a rotting body on the bed. It didn't reek but his skin ran freeroam of orange and yellow infection points. Some even a charred black for some reason. What's worse is that it was surrounded by the same flowers that were everywhere, growing out of his chest, head, and eye sockets.

I could have lived the same life without seeing that.

Could there be others? *More?*

Why can't I *smell* the body? Or bodies, if anything.

The overgrowing person looked like an older man and the roots grew through his beard and rough hands.

Out of curiosity, I looked at the door to inspect his chart on a clipboard hanging via a small white string. I grab it and start reading.

Two minutes later, I learned his name was Harold. He was born in the 1990s. He is 45, and he was a firefighter. Diagnosis showed he was trapped in a burning building for too long. He had multiple grade 3 burns and clogged lungs. He was apparently blinded in one eye, too. His pronounced death was exactly three weeks ago.

What a hero. Ordinary people like this don't deserve this fate, yet they still get it.

I gave a quick salute and left the room. Come to think of it, maybe it would have been better if I did see him. I may be the last person to do so.

I left and finally reached the end of the hall. Reaching to a giant door. It stranded a solid eight by eight feet. There were no windows, only the solid light brown wood that covered the whole door. Though there were steel door handles, next to one was a screen.

I tried pushing the door open, but it was solid shut. Roots and flowers peeked through the cracks and spread on the walls and floor. Even prying from where the roots seeped inside was getting nowhere.

Finally, out of curiosity. I swiped my hand on the screen, dusting it off.

Access Granted. Unit ID. "Family."

A robotic woman's voice yells. This caused me to jump, and then I heard a couple of low hums before the door opened a crack.

Woah. is this thing broken? Why would I have access to it?

I go push the door and try to fit inside. The door gets caught on something, yet I'm able to force myself through to the room of greenage. All of the room was overgrown, the roof sprawling with vines, and it all reached this smaller door that showed a miniature room, observable at all angles from this outside room. A vacuum-sealed door is shown ahead. The only entrance inside.

This outside room had windows that peeped inside. Tables behind them with overgrown roots and vines sprouting through the keyboards and cases of them. Like the lobby, papers littered through the floor. Though, they were different. All about "Tests" And "Classification" and whatever other big words that bolded on them. Reminded me of a rainforest cafe.

Unlike a rainforest cafe, though, these flowers were all the *same.*

Yeah, they were different colours. But taking a really close look at them, all flowers had five pedals, each one looking crumpled. They had three stamens poking out the centre; they were a matching white-to-pink ratio. And they were small, about the size of a large bottle cap each. They grew around the height of a foot and the stem colouring was a nice bank green.

I step away from the flowers and go to a monitor. I'm curious if these things will even turn on.

Who really knows until you know?

Moving a stack of paper away from the power button and clicking it did nothing. Bummer.

Though the papers were interesting. I scanned over the first page of the stack and read to myself the printed writing.

Day 1471

April 27th. 2035

Statistics. Jack displays a complete loss of control over his unnatural ability. Containment is scheduled for 0-48 hours. Location? Iron Pulse Industries of the ECTO organization. Headquarters at Loxahatchee, Florida. We can't make time for any further testing here. Our equipment is not authorized, and our whitelisted doctors are getting tired of ignored answered questions-

Can they make their documents any longer? Like, I get the specifications, but jeez.

Skipping random gibberish about plants…

Overall, our specialists are able to handle and contain the situation for ado times. We requested the Iron Plus side to make room for our subject. If they saw how unique this one was, they would accept it gracefully. We keep to the contagious note. Whatever Jack has, if the world gets anything like such. It could very much mean the end of it. All of the resulting answers to today's tests are below the signature.

Lead researcher.

Dr Laura Silvers.

And that, folks, was just the *Top page.*

That freaked me out already. So, imagine how I felt when the stack I had in my hand was just over 1400 pages long. Each one neatly stacked. And each one having a day and date. And all of them with the same printed signature.

I grab the pages, which, *thankfully*. Had 3 rings for binder holders. And put it under my arm, still holding my gun slightly.

Something slipped in between the papers. Looking over. I realized it was a picture.

I kneeled down and put down the book of papers.

Wait. aren't all books "The Book of Papers..?"

Anyways, with that, I picked it up and flipped it over.

In the picture. You could see a boy lying down in a gray bed, curled up and sleeping. He was in a white shirt and jeans. But that's all. He was a silvery blond. He has a pure face. No scars, freckles, nothing. Just…

Jack.

But that's not all the photos shown. Under his white shirt, through some weaves of his messy long hair, stems and rooting of flowers that sprouted on top. Each one is the same as I have seen around. The weird thing was it didn't seem to be affecting Jack. He seemed at peace.

"Excuse me?"

I whip around to look. And pause. My M27 in my hand, muzzle pointed down and the photo slipped through my fingertips because what I saw shocked me. And it *still* does to this day.

It was a girl.

September 11th, 2035

6:45 AM

Arlington Medical Research Facility. Texas

"Hello, do you know who I am?"

Is this a trap? I dropped the papers and raised my gun.
"Do you like flowers?"

I lowered my barrel down as quickly as I raised it. Yet kept the gun slightly up as I inspected her.

She was white. Hell, to put her in perspective, she looked like a pale supermodel. She was in a ripped-up coat and shirt with long-sleeve scrubs. She looked as if she was in her teens, maybe my age even. Her hair was white with silver tips styled in a wolf cut. Long enough to reach her chest. Which I am not going to describe too much in detail.

Let's just say *curvy*.

But her eyes.

Wow, her eyes.

Her eyes were big, intelligent looking. The opposite of oriental eyes — doe eyes of coal black. Did I mention her eyes are… beautiful? I was lost in her universal gaze for a moment.

Hypnotized.

But only for a moment.

"Excuse me~," she says, waving her hand.

"Oh… hello?" I quickly spit.

"Hello, friend. Do you know who I am?" she asks, twisting her head.

I start to sputter out, "Wh- what..? I mean- like… should I know you?"

"I don't know. I don't know myself either," she answers, ending with a "Mhm."

"You don't know who you are?" I ask, gathering my thoughts.

She shakes her head in response. She then looked past me and at the flora that grew all over the floor. "Do you like flowers?"

"I- they scare me…" I speak. "Did you plant them?" I ask, thinking back to the fireman I saw a few minutes ago.

She shakes her head again. "They were here when I awoke."

"When you… woke up?"

"Mhm," she says, nodding her head. She walks up to me and looks up for a second. She is 5'6 or 5'7. Yet she was a little skinny, even if she had a good figure.

She grabs my arm and walks me to the edge of the room.

"Pick flowers with me!"

"I- I'm sorry… I can't… When did you wake up?"

"Well… maybe a day? There isn't a clock inside here, and it's been cloudy out. So, the window is no help. It's hard to… to say…"

Then, her eyelids began to flutter. While her pupils rolled back, and she began to collapse.

Instinctively, I dropped my gun and caught her. Her head rolled around and settled on its side. There was a number on the back of her neck — "1" in black ink. She obviously fainted, but why? And from *what?*

I checked her pulse, and it read slow, but it was there. *She is fatigued*, I thought.

So, I grab her and hoist her on one shoulder. Still carrying my gun, I bend down and manage to barely grab the stack of papers with my fingertips, shoving them under my shoulder.

So now I'm marching my way out, pushing past the fancy door and moving past the flower-filled hall. Making it past the map, I get to the emergency lobby. And I see Ivys standing by the entrance. Her arms crossed and pistol in hand.

"Ivys! Ivys!"

Her stance freezes for a moment before she rushes in, pistol in hand, ready.

"Private? Who is this?" she asks, a stunned tone fills her voice.

"I- I don't know, she didn't know either-"

"Excuse me?" she asks. This time in a voice of confusion as if I spoke Chinese.

"Just- she needs help. She is alive but fainted from fatigue. I don't have any ammonia salts or anything useful in my bag to wake her."

"Get in. I'll drive. Your bag is already inside," she says while going around my new, refurbished, and robust windshield.

I open the backseat and drop the papers inside, then I sit her down in the seat behind the passenger side and put in her shoulder straps. I throw my M27 under the seat and Buckle her shoulder straps in.

"*Gasp!* Nasturtium…" she said, jerking awake. Before she fainted again.

I check her pulse again and…

Wait, no.

Not fainted. *Sleeping.*

Her pulse was a little faster. This time it was more powerful — the strumming was a recognizable rhythm.

I rest her head on the side of her shoulder and close the door. I then rushed to the passenger side as Ivys screeched the tires and jerked the machine up when I stepped in.

"She is okay. She is just enervated."

"From?" Ivys asks, not looking away from the barely visible windshield.

"She said she 'woke up' not too long ago. Yet she said that she didn't know who she was…"

"Amnesia," she says.

"Possibly ma'am."

"Listen, it's fine. She just needs to eat to clear up her memory. She is all skin and bones right now." She reassures me.

I nod in response as we drive the ruined path back.

"When we get back, we'll put her in the civilian camp at least after she recovers. See if you can find her parents while you're there. You're probably in the same age range as her, so she is under your care."

Great. Babysitting.

At least her parents will be relieved.

"Yes, ma'am."

"Right. To catch you up, we are leaving tomorrow."

"The new lady came to her senses?" I snorted.

"Seems Colonel and her made an agreement."

"And such agreement is *what* exactly?"

"1/3rds of her soldiers, and some of ours, will take the civilians to our base, AKA the school. While the other 2/3rds of hers will be allying with us on our trip to Houston. Which has the remaining access key start for our beacon signal. The amber alerts final access ligament is in the naval base of Houston."

With how far Houston was, combined with the time it will take to search most cities along the way, combined with our troop and cargo, this mission could take well over a month to complete —especially with the sentinels roaming.

"How far did I walk?" I ask.

"Not far, we should be pulling up now."

As if she made it appear, I see the wall of Humvees in the very limited distance, the crumbled buildings on each side of them, darkened and mysterious.

"Pretty, isn't it?"

Fun fact: That was not Ivys who said that.

I turn around to see the girl looking out the window. I wish I could agree with her, but my view was cracked glass and a book slot-sized peephole, yet… I still saw what she meant.

Smoke.

Ruin.

Ash.

Those are the only words I can use to describe the sight barely. Beautiful yet ugly, peaceful yet scary.

Unique. Yet *everywhere.*

Nothing fits it in a way where one word can describe it. Not one sentence, not even a *novel,* can come close to showing this hideously beautiful art piece of disaster.

"Have your flare?"

"Yes, First Sgt., its-" but when I felt my pocket, it wasn't there. "-its umm… let me check my bag."

She stops the car about a good acre away from the wall. And I get out and go to the back door. Opening it, I see the girl with my bag in her arms, hugging it.

"It's very heavy, you're so strong!" she says, letting go of the bag.

"Um… thank you…" I say, hiding my stupid, warm blush.

I grabbed my bag from her and put it on the street floor. Looking inside, pushing past the MRE, which caused my stomach to rumble again.

"You're hungry," the girl said after hearing it.

"I'm fine," I answer before loading the red flare and pointing the flare gun up.

Poomph!

As the sky streaked red, I put the flare back and placed my bag under the seat.

"Please eat soon," she whispered, seemingly unmoved by the loud bang of the flare.

I nodded and closed her door. Getting back in the passenger side, I painfully try to watch the Humvees move out of the way through the slit.

"I'm removing this "Windshield" the first chance I get." I scoff at Ivys.

She snickered. "It's your truck."

"Humvee."

"You get the point, at ease."

We make it inside the compound. As soon as we entered, we realized something was off. A lot of movement was happening. As we pulled in and parked, I saw panic, fear, and worry painted on faces. I get out and see Dominic. He's loading people, civilians, onto one of the three cargo trucks lined up.

I ran up to him, pushing past the people in the messy, disorganized line.

"Colonel?"

"Change of plans, Beast is on the prowl! We are leaving!" he shouts over the voices, loud enough for me to hear.

"Today?"

"Not today! *Now!* With our new allies!" he says, pointing at some of Zu's soldiers who ran by. They each had a Maroon ripped piece of cloth tied to their thighs.

As for Zu, she was loading up the truck in the front. She had a maroon-red band tied around her neck.

Allies.

Oh fuck, I forgot about the girl.

I rush back to my truck-

Ahem, *Humvee* and open her door. She, again, was holding my bag in her arms before letting go.

I take my bag and unbuckle her straps.

"I need you too, come on."

"Make sure to eat."

"Just come on," I say, grabbing her hand gently and walking her toward the trucks.

"GET DOWN!"

Suddenly, bullets rip through the air, like angry hornets, as they fly past, narrowly missing their quarry. Our soldiers push the assault back as the civilians rush towards the trucks.

"Fuck! My gun!" I yelled as it dawned on me it was in the truck. I mean- Humvee. Shit.

"Come on!" I yelled.

As the girl and I ran back to the car I threw the door open, shoving the girl inside before sprinting to the driver's seat. Before I was able to start the truck, the passenger side door flew open, and Dominic jumped in, with his polished colt drawn.

"All cargo units! Move out back to your destination! Now! 2 Humvee Units Follow behind! Take a damn growler too!"

"But sir..." a voice came up on the radio. One I haven't recognized yet. **"That's more than thirty soldiers, maybe if-"**

"I will not leave those civilians in the hands of Beast or his Sentinels! As of right now, our army is halved. The new frontier units will also bring an advantage! The gear they-"

The windshield... ish, anyway — it started deflecting the bullets flying towards it. The impact sound echoed all inside the Humvee, ringing our ears.

"MOVE OUT! NOW!" Colonel yelled.

I don't know if he was talking to the radio or me. But I didn't stop and ask.

I shifted to drive and stomped on the gas pedal. The truck- Humvee... the Humvee jerked forward, and I barely could see through the peephole, so it surprised me when I drove back to where the cargo trucks were originally parked, that's when I got stopped by someone.

Sgt Ivys opened the backdoor and rushed in, slamming the steel door.

"They are flanking the north quadrant by hitting one of the east side buildings."

Colonel, still holding the radio, starts yelling into it again as random bullets ricocheted off our Humvee.

"Cruiser one! Artillery one! Check!"

"Check!" Ijins voice yelled, recognizable

"Check!" said a voice I didn't recognize about ten seconds later.

"Status report!"

"Currently on the eastern side, defending the push of Sentinels," Ijin said, seemingly with a calm side. His tone was easy and understanding.

"Why do we need a tank there? It's just footmen." the colonel asked, puzzled.

"Negative. Units on motorcycles are pushing as footmen hold the flank, along with a machine gun mounted Toyota Camry is spraying us. Two rocket launchers were spotted, too."

"Damn!" Colonel sighs

As if on cue, an explosion echoes down from the east side.

"Damn!" he yells again, then brings the radio to his lips.

"Cruiser four and five! Meet at the northside building!"

"Copy."

They both said one after another.

He puts the microphone back and sighs.

"Let's see how good this frontier army is under pressure."

September 11th, 2035

7:13 AM

Frontier Camp, Worth. Texas

We parked in front of a building that had a sign over it, plastered on. It read "Allstate insurance" in navy blue text.

"On me," Colonel says, stepping out.

I grab my M27 and take off the safety.

"Not you, Private. Did you forget about something?" he says, tilting his head towards the backseat.

I look back and see Ivys and the-

Ah, right.

"Watch her." Ivys commands, "And eat. She told me you needed to."

"When an assault is happening right in front of us?" I castigate.

"Yes, eat," she says, severely harsher, staring me down before stepping out and slamming the door.

"Yes, ma'am." I sighed to no one, putting on the safety switch on my gun again.

Colonel now stepped out and followed.

God, I keep being treated like a child, all because of a stupid loss.

"Please eat," the girl says from next to me.

"I will." I sigh. "You're hungry too, so just- let's share a meal," I say, pulling up my bag from the backseat, taking out an MRE, and holding it out to her.

"It may not be warm, but it's really good."

She opens it and takes a small nip of the egg and cheese mixture. After a second, she smiles and takes a huge bite.

"Hey- Easy! You're gonna choke…"

I pull out the water bottle and the canteen I never used, fill it up, and give it to her. She chugs a little down.

"Thank you. Sorry," she says.

"It's okay, really. I'm glad you're eating." I say, opening my MRE, taking a nice bite of the "Spinach Beef Mix"

As I watch her eat, I begin to notice more about her. Her cheeks were hollow, yet a faint natural blush was starting to come back. She wore a blue-ish windbreaker, which, on her left side, a sticking of a brand was in her pocket. It read in bold text, *"Iron Pulse Industries."*

Right. I know them. It's the company that worked for ECTO. They were a breakoff company of NASA. Their job? Well, send trash to space, while mining out the asteroid belt for our precious metals. Gold, magnetite, nickel. While ECTO worked on that, IPS (Iron Pulse Industries) — hired PMC, made weapons and protected ECTO's facilities nationwide. And they have *serious* firepower. In fact, they were the ones who invented the F-25 Falcons.

**Ting! Tac! **

Two gunshots hit the Humvee… somewhere, so I grab my bag and throw it on, as well as my M27, and, once again, I take off to safety.

"Stay here," I say, opening my door.

"What's your name?" she sits up, inspecting me.

"Will…" I answered before getting out and closing the door, running around to the back of the truck.

After about a minute, I stalked my way to another Humvee closer to the bank. Something hit my Humvee, whatever it was, it was close.

Next to the building I made out was on the left of the alley. It was nothing special, and I almost chose to ignore it. Until I saw three men walking towards me down it. The Humvee's Driver-side door was facing the alley about eight feet away. I peek over the hood.

Sentinels. Armed Sentinels.

I'm in position for an easy hit on them if I mount my gun on the hood… I could do this. I place my gun on the reinforced body of the hood, finding a spot to keep it kinda sturdy between two armor plate welds. Pointing the barrel down the alley, I readied myself.

They haven't seen me yet. Good.

I breathe in, filling my lungs with calm, dirty air.

Then I pause.

Two of the frontier soldiers came out. Each one was walking in front of the alley.

"NO! DON-"

The shirtless sentinels saw them and shot them dead.

Dead.

I wanted to cry.

I wanted to vomit.

I wanted to run.

The men down the alley yell something, and his fingerless gloved hand points right in my direction. I managed to crouch down before a spray of bullets whizzed to where I stood, hitting the Humvee and anything around it.

Then it stops after about a minute. The aura silent all around.

I get my gun back on the hood,

Aim.

And fire down the alley.

The whole pass was just my bullets showering the entire alley. One of the sentinels took cover behind an iron dumpster while the other two were gunned down.

All I heard were the terrifying pops of my gun. All of them scared me.

Click!

Ammo.

I duck down and drop my mag. I throw my bag off and kneel down to search for another mag.

"He's over here!" said an unfamiliar, strained voice.

Damn! Reinforcements!

Their footsteps got closer. Heavy stomps were now all I focused on as I desperately searched for my spare magazine.

Their loud stomps stopped, and they stopped *very* close.

I throw down my M27 and pull off my pistol, listening closely.

Faintly, I can hear their boots crush the asphalt and dirt slowly, each step gaining location on me.

Finally, I see one of the men's Nike shoes take a step to where I can spot it. So, I aim.

And shoot the front of his creased shoe.

"AGH!"

He cries out, bending forward to where I see his full top half. So, I shot him in the head.

I hear a loud stomp behind me, and I flip around to see a white, skinny sentinel sneaking up on me with a knife. When he sees me spot him, he charges, but not before I get up and shoot him twice in the chest.

I felt a kick to my gun as my arm slammed into the Humvee. Releasing my grip, I spin around to see a third man. The one that hid behind the dumpster and survived. He was now pulling out a six-inch knife with his left hand, lunging at my back.

I duck and punch his genitals when he stumbles back, yelling. I grab the arm holding the blade. He comes down and starts wrestling me for the knife.

I slam my elbow into his throat and finally grab the handle of the blade he loosened his grip on. He manages to flip on top of me and grab my throat. I slash the blade once, cutting his left wrist. When he lets go, I turn him around and pull down his right arm while I wrap my legs around his waist. I aim the blade as he struggles to get away from my grip .

and stab it deep inside his throat.

His blood, warm and thick —trickled down his neck, getting all over my hands, all over my shirt, and down the left side of my cargo pants. As he tried to talk, to scream, all he could do was cause more of it to leak out. He spits a mouthful of blood one final time. His body slows in breath and twitches as he dies to my hands, Still on the handle.

My thoughts were blank as I pushed him off of me. Even when I stood up, and stars filled my head, no thoughts did.

I killed this man.

I killed everyone here, these people. Even the blood of frontier soldiers was on my hands. If I just shot the damn goons beforehand, if I didn't think so slowly and reacted even slower, they would've been alive.

And yet, my emotion was still blank.

I grab the bloody shirt collar of the now pale man and move him over. Grabbing my dirty pistol and dropping the mag, I go back to my open bag and pull out a box magazine, along with the M27 magazine. It had the letters STANAG on the bottom. I shoved the box magazine inside my M18 sig, holstering it back on my thigh. I go to the M27 that I threw next to the bag and pick it up, dusting it off and sticking the STANAG inside. Hearing the click, I cock back the charging hammer and watch it push a bullet into the deadly chamber. I put the safety on and looked up.

I know I'm blank from it now. It's most likely a late reaction. The shitty feeling will kick in any hour now.

The blood was cooling off quickly. I looked at my shirt as it was covered with sweat and blood. All of it was leaking down to my foot, staining my pants and a little of my ripped coat. The sight was *disgusting*. Whatever was inside this guy was all over *me*.

With that information, I grabbed the rear-view mirror and emptied half of my lunch out.

Fuck I'm a pussy.

I wipe my mouth and pull my head back up, the weight of myself barely bearable.

Dominic runs out from the bank, Desert Eagle in hand — spotting me.

"Will! Didn't I-"

Then he spots the first body by the Humvee Bumper. He walks over silently and comes around to see the sentinel I shot in the foot and, finally, the one I stabbed. Looking down at the brass casings and, finally, down the alley, I could see the four corpses. Two of them were our own soldiers.

"The frontiersman?" he questions

"Killed by the sentinels," I answer

"You took out five soldiers?"

"I took down five targets," I replied. "It was simply a fight or flight reaction."

"Good reaction, Private. You're serving to be a fit soldier to this army," he says, finished sticking the last round into his chamber. "Do you have… extra clothing?" he asks, staring at my blood-stained body.

"An extra OCP, but nothing else."

"We will find one for you," he says. "Just take it off. Rip a dry end and tie it around yourself."

So, I follow the command as the colonel goes back inside. The gunshots stopped so they probably were done. I had just finished tying a knot around my other thigh and leaving the loose strands to hang like flags when Monroe walked out, sweat-filled. He stared at me while I put on my coat with no shirt. Great, now I look like these stupid sentinels. Though, I'm not completely topless.

Monroe drops his bag and opens it, taking out a shirt and chucking it at my head, I pull it off and hold it out.

"Thanks," I say.

He grunts something and walks off.

I take off my coat again and throw on the shirt. It was a little loose, though I'll grow into it.

"Hey, kid?" Monroe's voice simmers,

I look up as I'm putting on my coat to see him at the same place, only staring at the man I shot with my handgun.

"Do you have a jetted pocket on your coat?"

I knew I did. I just never used it because I had four pockets already. The last time I used the inside pocket of the jacket was when I was stealing.

Yeah I stole. I know, I'm a thief. But I just… I was stupid for it. I didn't steal things that were even in value. Pencils, hair ties, buttons… I really only stole to see how good I could be at stealing, how much I could get away with. And that hasn't ended, sadly. The last thing I stole was before the meteors. It was a basic Army pen-off from the Sgt at my school. But after the whole… Kabooms, I never really had a thought of stealing. Too worried about many, many other things.

"Yes, I do."

"Put your mags in there. And do it faster than digging through your bag, you can get yourself killed doing that.

"Oh- yeah… smart," I say, surprised.

Was he helping me?

Why was he helping me?

Is he being friendly, or is he planning something? All different types of questions scrambled my mind as I shoved two M27 mags into the left jet. Though one was empty. But on the right jet, I put three mags for my pistol. While I was pulling my hand out, the tag of the jacket inside the pocket gave me a nick. Papercut.

The weird thing about that is the bleeding stopped almost instantly. It was sealed in under a minute.

But it was not bleeding, so I shook it off, picked up my bag, and went back to the Humvee. Looking at the stupid-looking steel door window.

Yes, it's safe to have, but it's just stupid-looking, you know?

I open the rear passenger door and throw my bag inside. The girl was on the other side. She was sleeping against the door, silent and calm.

And I'm like fucking, how?! Bullets are being shot right outside of the Humvee. Some hitting the armor! Did she forget her safety settings, too?!

I close the door and make my way to the driver's side. I get in and strap in as everyone starts to come out.

Ivys was the first one to the back door. Opening it and seeing the girl sleeping soundly.

"You're dealing with this private. Get back here. I'm driving."

"What- but…"

"That's an order, Private!" she says loudly.

"Okay… yes, ma'am."

It's not like I'm going to drive well with that window anyway. So, I got out and opened the rear left side door that woke her up.

"Mmh… oh- did I fall asleep?" she says in almost a whisper. "I'm sorry, Will…"

"It's okay, watch yourself. I'm going to sit in the middle," I say as she lowers herself, and I squeeze past her. Preparing myself for another long ride.

September 11th, 2035

8:45 AM

Plant City, Worth. Texas

"NO NO NOOO!!!" Trevor frantically bawled. He clawed at his hair and the air around him, yelling slurs at the top of his lungs.

"PULL THE MEN BACK! SOMEONE! GO! NOW!!!"

Three soldiers around the roof ran off, leaving Trevor… seemingly alone.

"Are you sure you don't want the sponsor?" A deep, dark voice filled the man's head. It came from everywhere as if he was a ghost.

"Fuck off! I know what I'm doing!!!"

"Clearly, they know, too. You are able to see that, right?"

"Shut up! Shut up! Shut up!" Trevor yelled before running up to a brick wall cover and smashing his head into it.

"Sir..?"

"WHAT?!" Trevor's stentorian voice rips, his hair a huge mess, his jaw clenched, and his deep purple eyes full of rage.

"W- we pulled out, sir… we seen—"

"Load up the rest of what we have! Everything!" Trevor commands.

"Sir? Including the—"

"Yes! The tanks! The trucks! The bikes! Everything!!!"

"Yes, sir!" The soldier in front said before running off with three soldiers.

"Give me space, beast, I can make this army an unstoppable force."

"What?!" Beast yelled.

The ghostly voice went silent.

"Speak voice!!!"

"Just give me your permission. Let me bring you supplies and in return, you follow my orders, and your first order is to bring me the Seed. Doesn't that seem easy?"

"Sir." Bush said to Trevor, bending down, itching his head.

"What maggot?!" Trevor yells again.

"What about the captured civilians?"

"Tie them up! Bring them too! We're going back to home base! Don't kill any yet! I need to set an example of them to Dominic!"

"Yes, sir," he quivers, running off, leaving Trevor… almost alone.

"I'll be expecting you, Trevor," the voice sang.

"Ugh. I need to eat, voices in my head are annoying!" Trevor whispered angrily.

September 11th, 2035

1:00 PM

IH-45 Road. Texas

"When we park, Will. I need to talk to you," Dominic says.

"Yes, sir." I sigh; I have two wild superstitions about what it could be on. It's either what happened before or during the assault.

Or what's happening now, while the girl whom we saved, sleeps soundly on my shoulder. Even the bumpy road could not phase her. And you have no idea how tense I was. On my right side, Monroe was sitting up, calm. His brown eyes followed forward. Ivys, who was clearly used to driving anything with a book slot as a window, was calmly talking to Dominic, who was looking at a map of Texas in his hand.

We are now leading all six Cruisers, two Growlers, and three Artillery units (one a tank, while the other two are Mounted Humvees, one being a 50 Cal. Machine gun, and the other being a grenade launcher.)

The three cargo trucks we originally had took the rest of the cruisers, some being frontier soldiers and some being rebels. We may have taken two bikes from the skirmish — you can't blame us. They had side carts with gun mounts on them, and they could fit three soldiers each. That leaves about forty of them and twenty of us, eight Humvees, two bikes, and a tank.

Seems like an army to me.

Oh, and one more addition: a civilian — which is currently sleeping on my shoulder.

"ETA?" Dominic asks Ivy.

"Corsicana is another hour, sir."

"We've already been on the road for an hour," he scoffs.

Great, The ruined road trip. The sequel.

"Feeling hot yet?"

I looked at Monroe after he said that. He sat the same, his eyes not peering away from the front.

"Is my face turning red?"

"No, but your neck is sweating like a pig."

"Pfft," I blurted, "Pigs don't sweat."

"What? Yes, they do!" he comes back.

"No, they don't, bro. Look it up."

"Even if I could, I know pigs can sweat."

"When have you ever seen a pig sweat?" I snap back as I grab my old OCP from under the seat and wipe my neck with it, making sure to avoid the bloody parts.

"There is one sweating next to me now."

I muffle a laugh. "And besides that?"

He pauses and looks up, thinking.

"Come to think of it, I don't think I have…"

"Exactly. Fact check yourself."

"Maybe it's better if I don't," he says, looking down. "If I ever see another pig, I'm making pork chops on sight."

"Yeah, right…"

Weird enough, that reminded me of this.

What am I doing?

I'm serving in the Military not for myself and not by choice but for survival. This is not a battle over politics or disputes but one thing we all crave — to survive.

There is no other choice but to think of my father whether he was here. What would he be doing in this situation? It's so bizarre that nothing came to warn us about this event. My father was a soldier, a real soldier. He knew what he was going to do. But the… "soldiers" today are just common people, neighbors, friends. All fighting in a dying world.

My father still would have kept hope. He would have said something like, "Relax! What matters is that we are alive. And guess what? We can fucking smile because no one else would!" Then, he would smile his brilliant chiseled cheek smile, his face as similar to mine but aged. A blond stubble combined with a soul patch. I can't blame my mom for swooning over him. He reminded me of a pirate sometimes.

"I want to say I'm a little disappointed you held out on your skills like that."

I looked at Monroe, who looked back at me with an serious expression crossed on his face.

"What do you mean?"

"Don't be a humble kid! Five kills? One by a knife to the throat? You had a kill more than our colonel! When you wrestled the Frontiersman, you looked vulnerable and weak. I'm shocked, honestly.

"In reality, I did it to survive, and that's really it," I say firmly. I was not going to describe that talk with him.

"Rather so. For a kid, you're not a half-bad soldier."

Is he mocking me?

"Why are you- like- being "buddy-buddy" with me now? Is there something you want?" I snap back. God, this guy can make you feel bi-polar.

"Honestly, I fuck with you when we were at the camp because you seemed puny. Though, you're serving this troop like no other private I have ever seen," he said.

"I thought you were the meathead here," I say as I let out a sigh. "Wait, you still ruined my car with spray paint!"

"I actually made it better."

Okay, now he is mocking me.

"You fucking prick!" I hissed.

"If the colonel had seen you, he would have left you. Plus, you made it stand out way too much. Once you joined this army, your car became our car."

I'm glad the Generals couldn't hear us. Not only was the ride loud, but they were in their own conversation.

"Okay, but what do you mean the colonel would have left me?"

"Colonel is… to the word, you know? He says do this, then do what he says in the most literal way possible. You were there when he said "no special designs," so you should know if he saw it, you would have been ditched.

So, what I'm getting at is that he saved me?

I was not expecting even to come close to liking this guy. Like, at all. But right now, he's seemingly cool.

"Thanks…" is all I can say.

"Mhm." That is all he says.

My mind was locked. I just thought this guy was an asshole. Instead, he is a… friendly asshole. His comments and looks though, they had a different story. So, anyway, the rest of the drive was physically silent. But my mind again was loud and annoying.

The girl, around twenty minutes later, pushes herself more into my arm. Crushing it with her chest.

I don't know what to feel.

Practically, the rest of the trip was this girl cutting off blood flow from my arm when we reached the outskirts of a city. Another five minutes of driving, we pull up to another abandoned lot. Not lots of rubble around so we all parked there.

The girl groans as she wakes up from the sudden stop. We finished finding a spot for us to park, and I saw soldiers already packing down sleeping bags for the night.

"Are we home yet?" the girl asks.

"Not yet," I answer.

"Everyone out. First Sgt, take the civilian with you."

"Yes, sir," she replies. Stepping out with Monroe. She comes around and opens the girl's door.

"Come with me," she demands

"Okay," she says, not caring for Ivys tone. She looks at me after she pulls herself out.

"Goodbye, Will," she serenely says, wiggling her fingers in a wave.

"I'll be out there soon," I say, nodding back before Ivys closes the door.

And here I was, alone again with the Colonel.

7:01 PM

"Do you know why I called you in here?"

Ah, he's doing a callback from our first time.

How lovely!

"No sir."

"Here, gonna make this simple, the civilian. You are to take care of her and nothing more. Which means no feelings."

"Feelings?" I ask

"It's bad enough she is even with us. I know why she had to come, but she cannot distract you until we are able to split more troops back to base, you are to take care of her. No feelings."

"What does "no feelings" mean, sir?" I ask. I think I already know the answer, but I'm not 100% sure. The thought never even came to mind.

Me? Liking her romantically? I have way too much to worry about. My brother is my first priority. The thought alone is kinda degrading. She lost her memory. It would be like I'm using her.

"It means don't get distracted with her. She will leave with our next refugee surge. Got it?"

"Yes, sir," I say, my tone firm.

"Oh- and good job today." He says in a more relaxed tone. "Your tactical self helped save many units. Men like you send… beast… running for the hills."

I noticed he hesitated, saying, "Beast."

But I was more focused on how I didn't deserve this gratitude.

"I couldn't help those two Frontiersmen during the fight." I sigh.

"It's a law that people will die in war. And sometimes you can help prevent it, while other times you can't."

"Sir, it's still my fault."

"And that still doesn't matter. You weren't the one who shot them down. What matters is that we move past it," he says, "and kill… the beast."

"Finally, enough of that "prosecute" bull you said during the school ground meeting," I snorted.

"At ease!" He yells, making me jump.

"Fall out and rest up. Don't change my mind or you being here, you're young and healthy. You had your whole life ahead of you, but you're tucking in with us now."

"Yes, sir. Thank you, sir," I said as I got out.

As time passes, I make myself an area to sleep. Clearly, news travels fast. I was getting all kinds of praise. I didn't feel any of it I deserved. So, I went to sleep, sulking, emotionally defeated.

September 12th, 2035

7:00 AM

IH-45 road. Texas

And so, the next day was just a resupply of water and gas. We managed to get a hold of the girl. She still had her amnesia, and I still had my grief.

Anyways, we were back on the road around 10:00 am or so. Driving through the city, still on the path.

"Is that it?" Ivys says from the steering wheel, looking forward as if she saw a wooden shack.

I twist my head to the side a little and peek at a… wall.

It was not an actual wall, though, but a wall of tanks. Ivys wasn't in disappointment but in shock. Five tanks lined side by side, their muzzles pointed right at us.

Dominic grabbed the microphone of the radio and pulled it to his jaw. "Halt, men! Don't move an inch!" He yelled.

Another voice filled the radio, cracked but understandable.

"Army over the horizon. State your name and intentions, or we will open fire."

Before Dominic could answer, another voice cut him off.

"This is Commander Zu of the new frontier. I order you to allow me and these men to pass inside."

"Copy that Commander."

Two of the lined-up tanks started to move out of the formation, and behind them, people. Regular civilians dressed up in heavy blankets and scarves were walking past, looking through the entrance, curious.

So, as we drove in, we could see a huge army. Not a world war large, but enough to be classified as powerful. Lots of units with them. I can see men and women welding tanks, Loading rocket launchers, and even a couple moving a Heli from a hanger. These guys were stacked. We were in the heart of this army.

Not only that, I managed to spot some snipers peeking on the inside from four high vantage points. This is no camp but a fully functional base. It was in the middle of a cross-section of the street, and four buildings provided some cover over the corners. In the middle clearings, in front of the buildings, stands were shown with canned foods, knives and guns were shown to us as we drove in.

This place was a heavenly sanctuary. We were led to one corner building, which was a parking garage they let us use. The brutalist design shows how formidable it can be.

Ivys takes the microphone that the Colonel was still holding and started talking.

"All rebel forces. Fall in front of the colonel."

And then we got out. A few minutes later, we were marching out of the parking garage entrance. Our half of the army was in a formation, while Zu's army wasn't. We continued on until Dominic halted us and made us fall in a formation right in front of a short Indian man, maybe late 40s, with tampered skin and a tanned face. He was accompanied by four well and heavily geared soldiers. They were all dressed in black and brown.

To shorten the whole complicated hour that passed while we stood there, he's basically the mayor. And all he really did was explain the rules of the place, no fighting, stealing, etc., while also promising to supply us with gear and stuff, basically a whole lot of yada yada.

It ended with us falling out and calmly being relieved of our duties for now. The commanders and the mayor afterwards started walking in another direction while the army split off. We were told to meet back at the parking garage at 6:00.

Free time is nice to have. I only had one condition set by Ivys.

2:23 PM

"Are there any pretty flowers around?"

"No… I don't think so," I tell her. As we walked past the stands, my hand clutched softly to hers. Right now, I'm looking for ammo. I need some more for my M27 because that thing eats ammo quickly. I had it strapped to my back to see if anyone would at least walk up to me and help me find the ammo I was looking for. Plus, I needed mags for it, too.

"Hey, umm… girl. We should… find you a name. A temporary one, at least until you gain your memories. It's kinda rude to just call you… girl."

"Yes!" She says softly. "That would be lovely! Mhm!"

"Okay, well… you do like flowers, so how about we name you after a flower?" I suggested. "You like flowers, right?"

"I love flowers! There are so many! Maybe I can be named Helianthus annuus L! Or Rosa Rubiginosa! Maybe even Bellis perenn-"

"Woah, woah, what? Those are flowers?" I question, trying to remember what she even just said.

"Mhm! Or there's Narcissus pseudonarcissus. Or Le Amaryllis. Or even-"

"Wait, what did you just say?" I cut her off.

"Amaryllis! A flora plant consisting of white, pink, or even red, they are a part of the genus flower family!"

"I know about that flower, my mother told me that they represent strength and determination. Why don't we go with that? Seems like a pretty name." I say as I hide my blush.

"I love it! Thank you Will! Mhm!" She says, jumping slightly and smiling her bright smile that almost shows the sun rise in my eyes.

I knew about the flower fact from when my father died. My mother only covered his casket in them, when I questioned why, that's when she taught me the meaning of them. She was showing how much bravery we saw from him. I aspire to be like him.

"Did you just... name her?" Says a familiar voice from behind us.

"Zu!"

"Commander Zu." She corrects. "You still haven't answered me. And what's with the number on her neck? Why not call her one?"

"Well... yes. But-"

"Ma'am, I lost my memory. I only remember certain things. This does not include my name. So, we have come up with a temporary name for me." She speaks. And she says it very professionally. As if she knew she was talking to someone of high power, it surprised me, and it definitely surprised Zu. But she simply nodded. "I see."

"Were you eavesdropping on us?" I ask, Continuing to walk.

"No, I just heard you before I saw you. So, I listened in." She says as she keeps pace with us.

Uhuh.

"I'm looking for rounds and mags," I say.

"Follow me. I don't think we have anyone-"

Suddenly, a man ran into her, knocking her a little back. He was big enough to do so. He stood as tall as her but was insanely built, thick skinned, white and scarred. His one unique trait was a dog collar he wore on his neck.

"Are you blind bitch?!" He says, shoving her shoulder.

So, I was about to step in.

Only before I did, she pulled out a handgun from her hilt and shot him in the thigh. He cried out and fell on the ground, Gripping his leg. That's when Zu started to point the gun at his head.

Instinctively. I jump in front of her and hold my hand out to the gun.

"Move, private."

"We can settle this a different way Commander." I plead. I stopped a second before I finished off with "Commander" so I hope she remembers her standpoint and thinks rationally.

"He showed the defaced aggressiveness that puts everyone here at risk. His actions led to this consequence. She says, trying to aim at the man groaning on the floor behind me.

"I'm pretty fucking sure he learned his lesson when you shot him Z- Commander." I correct myself before she thinks about shooting me next. "Just lower the gun..." I say in an easing voice as I put my palm on the barrel slowly and lower it even slower.

"Your commander will hear from this." She hisses, shoving the gun back where it was and turning around. Walking off.

"Fucking whore... OW!"

Turning around to see the man digging the bullet out with a pair of rusty tweezers. I wince myself as he pulls out the copper coated tip covered in blood.

"Will..?" Amaryllis says softly. And more quietly than usual.

"Huh- oh... yeah?" I answer quickly

"You need to treat him."

"What?!" I whisper back, "I don't know how to!"

"Grab a bullet, a full one," she says.

"Huh..?"

"Get a live round from your gun," she says.

I kneel down and flip my gun to my front, pulling and releasing the charging hammer while a 5.56 bullet comes out. I grab it from the floor and show it to her quickly.

"Good. Bite the copper tip off and pour the gunpowder inside onto his wound."

"What?!" I say, whispering... louder I guess

"It will work." She reassured me.

Sighing, I drop my gun and walk up to the guy, kneeling down next to him

"Stay calm. I'm going to help you." I say in the same easing voice I used on Zu.

"Ima kill that bitch!" He says, relaxing as he lowers himself, propping himself on his elbows.

I bite the copper tip, pop off the lid, and pour it on the wound.

"Wh... what are you doing?" He asks.

"Now, light it on fire," Amaryllis says as she kneels down next to me.

"Are you serious?!" I practically shout.

"Please?" she says, staring with her big, pretty eyes. "Trust me."

Holy shit. I can't win with her.

"I don't have a lighter."

"I do."

When I turn back to face the man, he's pulling out something from his pocket. He puts his palm out with a windproof lighter shining in it. Fancy and elegantly made. The cap looked as if a snake was wrapping around it. With green-painted eyes and golden skin, it wrapped around a Texas flag handle.

"Okay... bite something, sir."

"He sticks his shirt collar in his mouth and lights the gunpowder on fire.

The man holds back a yell from under his teeth as the fire quickly spreads. When it ends he tears up slightly with a long exhale.

"Now wrap the wound," Amaryllis said.

I didn't manage to mention it, but a small crowd of people gathered around to watch the whole ordeal. A woman from in it walked forward with a small roll of tan bandages in her hand. She handed it to me, and I managed to thank her. And so, I knelt back down with the man and snugly wrapped the bandage around his wound.

"The… the bleeding stopped." He said in between heavy breaths.

"You did it, Will," Amaryllis nodded.

"I swear to God's cross I'm gonna-"

"No!" I cut him off. "I can't save you twice."

"Whatever," he sighs as he gets up slowly, limping.

"I have actual medical stuff in the tent. Follow ahead," he said in a deep American accent.

"Yeah, we wish we could, but… we are currently on the hunt for ammo…"

"Then yet both of yer asses up then. That's my profession in the flesh. My name is Coal, current weapons smith and satans personal pain in the ass."

September 12th, 2035

4:06 PM

Corsicana, Texas

We spent the rest of that hour (plus some) talking to him at the tent. He treated himself correctly when we got there and he was walking a little better twenty minutes later.

Afterwards, I finally managed to do my shopping, asking him for two more spare magazines for my M27 and to fill back my empty one. He did while also lending me fifty extra 5.56 rounds in a case. And he told me to keep it all free of charge. He threw in the lighter for tax.

"What?"

"Keep it bubba. A keen thing to have a lighter always on yer body."

"It's way too nice, though," I complain; I know I'll ruin it due to my situation.

"You fucking twerp. Just take it!" He commanded, with no intention of taking it back. Clearly, he was the nice-aggressive type. His Canadian accent, mixed with his pirate tone, made him sound like the typical American stereotype. That's basically him, actually. He had a American trucker build.

Also, he had the dog collar on his neck.

Amaryllis asked him about it. We learned that it did have a dog as the original owner, but the dog sadly died under rubble. What's worse is he didn't get crushed but trapped. He was without food or water under thousands of pounds of concrete.

Coal for the next week fed and gave the dog as much as he could fit while also trying to move the concrete without crushing his beloved friend. The final problem came from limited food and fresh water. When he ran out, he decided to give the dog his rations. And when he ran out of those, he had to make the choice.

To give his partner a quick end with a gun rather than watch him starve to death slowly.

He wears the collar as a memorial.

A grim story. It was sad enough to have Amaryllis sniffling tears, holding my shoulder when we left.

5:48 PM

We were back to the parking garage and sitting by my Humvee. Well, I was, at least. Amaryllis was inside, sleeping.

At least she is feeling better.

Our crew of soldiers started rounding around back inside. I was busy loading and organizing my mags. Of course, I was keeping the extras in my bag but I decided to go with Monroe's idea and stick my magazines inside my coat. Three handgun mags on my right with a comfy M27 mag with them while the other three were on my left. Luckily I didn't forget my gun anywhere this time. Though I panicked for a quick second after I didn't see it. Turns out it was on my back. Though looking for it showed me the soldiers.

They used zip ties to connect two mags together. That was so smart. So, I do the same.

Well… I do get them to do it for me… after getting punched in the gut but hey. I'm alive. And with zip tied mags

And if you're wondering, no. Monroe wasn't the one that punched me. Some other meathead did. On the contrary though, Monroe punched him in the face afterwards.

No, no, not to defend me. He wanted zip ties, too. And that was his easy way of asking nicely.

Finally, a whole hour late, Dominic and Ivys came inside as the blotted sky started to dim down. Darker than it usually was, I could barely see Dominic's scowled face.

"At ease!" He shouts louder than usual. His voice bounced off the roof, putting us all in fear. No doubt, he was mad.

"Alright, men! We have good news and bad news!"

Clearly the bad news was really bad news. Even Amaryllis, even after being awoken, was watching and listening.

"So, these people are not moving from here. Not one will be allowed to come with us to the home base. What's also an issue is they won't take any civilians from us. Stated they don't have the food and accommodations for the civilians we choose to bring. And the ones we have brought already," he says, eyeing Amaryllis.

"The Sentinels attack this camp every now and so, the deal we were given was if we go smoke out the Beast from his home base, we have permission to bring whoever, wherever."

All the soldiers chattered murmurs. Equivocal on whether to feel happy, angry, scared, or ready to kill.

Maybe that one is an obvious one.

"For our good news!.." he yells.

Finally, Good news.

"We know where their home base is located!"

Oh god. That is not good news.

"They reside at the Coffield unit. Some of you may recognize that name, and it's indeed. The prison that I have sent Trevor to. Clearly, when the meteors took over, he used the structure for his own personal endeavours."

"Permission to speak, sir!" Yells Ijin. Raising his hand slightly as he leaned on his Humvee.

"Granted," he nods.

"Sir, I counted tanks, cars, bikes, Turrets, flares, and guns of all kinds. Can't they just do it themselves?"

"And that's the kicker." Dominic states. "They are willing to provide gear and supplies but as of now, they don't have any real trained troops, aside from the sixty or so New Frontier soldiers. Only twenty of them were actual Navy soldiers. Though one is a retired admiral. Which I respect dearly…"

Admiral! That caught most of my attention. The Admiral is the highest rank in the Navy. It has to be-

"Personally appraised by the United States president…" Amaryllis finishes my thought off from inside the car, to no one at all, seemingly.

How in the hell did she know that?

Your guess is as good as mine. Her statement made me question how she knew more about military knowledge than me if she forgot… everything else.

"At ease, men!" Ivys says, quieting down everyone.

"Permission to speak, sir?" another voice asks. It came from a young-looking man. He had short, gold hair and was standing a 6'1-6'2 frame. I recognized him to be one of the Tank drivers.

"Blaze?" Ivys nods

"I don't want to state the obvious, but can we not take all these tanks here and blast them to high hell?"

The men around him nodded in agreement and chanted hoots.

"The Sentinels captured around 200 prisoners. Each one is held hostage at the fort. Ninety per cent chance that the Beast is in there, too. As much as it would be worth destroying the base entirely, we need to save the civilians' lives.

Everyone stills a silence while Ijin raises his hand again.

"So, what's the plan?"

"We regroup with the Frontier army and discuss the plan from there. What's for sure is we will be with them in an assault tonight, retrieving everyone who is held hostage and leaving with low traces. It will be dark. There will be no moonlight or stars, making things impossibly darker. But that will be our advantage when we get there, making us very difficult to deal with. Everyone will be hands-on.

"Yes sir!" The soldiers yelled and ran off to their Humvees.

"Blaze, soot, get guns. We are not taking the tank. On the contrary, it's too slow and loud. Ijin, take them to find something that shoots."

"Yes, sir!" they yelled, running inside the Humvee they were next to.

"Ivys! Get a sniper. Meet us back outside the garage.

"Yes, sir," she says, moving away swiftly.

"Light! Go to the mechanic's factory quadrant and replace that damned windshield. Find a place for that civilian, too! She can't come."

"Yes, sir!" I salute, jumping into the driver seat of my Humvee.

"Everyone else! Follow me outside and park in a small line! The frontiersmen will take the rest and park with us. ¿Comprendes?"

"YES, SIR!" Everyone yelled

The sound of heavy engines engulfed the air, echoing and roaring through the parking garage. Suddenly, when I was about to pull out, my driver's door opened; Ivys stood there looking down inside.

"Want me to drive?"

6:10

In reality, she really just needed to get close to one of the watchtower buildings. Luckily for me, it was right by the mechanics. The window swap was quick and silent. I was so thankful that I could drive again.

Ew… I sound like… Alice.

Anyways… I picked up First Sgt, and she offered to watch Amaryllis. I couldn't find anyone that would take her. So, Ivys said she would watch her while taking shots, she would be safe and far away. But I mostly hope she looks somewhere away from the chaos.

So, I agreed. No one around was taking a "civilian" until we were done.

But yes! Finally! I managed to drive back to the troop with my new windshield!

Oh, with Ivys and Amaryllis, too, I guess.

So, I parked and fell into formation, splitting off from Ivys and amaryllis. They both went forward to the front of the army lineup. Next to Dominic and Zu.

Wait… Zu.

Holy shit! Zu! She's the ex-Admiral! It took me a while to realize that because, believe me. She was not the one I thought was an ex-Admiral. Maybe she took over for someone after a death or something. And where's her partner in crime? That blonde boy was nowhere to be found. I can't even remember his name.

"Forget fidgeting in a formation! Fall out, shut up, and listen up! so the plan-"

Okay. I don't mean to be annoying with sum-ups, but this was another long-overcomplicated speech. So please allow me to sum up. Most of the men will push a full-scale wall assault to the western side of the prison when we draw out the army and distract them from our small platoon that will be infiltrating the base from the east side. Rescuing as many people as they can, Ivys will be focused on Trevor. This was possibly to be the one that put down that rabid beast.

"-all ends tonight. A round is on me after we are done!"

That caused an uproar of cheers. I wonder which bar we will go to?

"Men, I want you to remember something. A soldier's promise passed down from me to you, men and women. When we get there, I will be the first person to set foot on the action field. And I cross to God that I will be the last one walking off that field."

"Yes, sir!"

"Let's go, hot rods!" Zu yelled. "Fire up your engines!"

September 13th, 2035

12:02 AM

???? ???, Texas

It felt hella nice to drive with a windshield again. I finally wasn't eating ash and dust from tires spitting them at me. The road that we took was impossible to see the name of. Dominic drove with no headlights and very slowly. But we were on it for about fifteen minutes before the radio echoed that we were ten "clicks" away from the base.

Whatever that means.

I stopped and dropped off Ivys and Amaryllis, waving them both goodbye and good luck. Two platoons will be pulling in on the other side on foot. We're gonna jack their prison buses and help everyone get out of that hell hole.

We started to see the prison's dim red radio tower lights a good 100 or so feet away. The engine shut off, and we jumbled together in strips, lining in into position.

and so, our grenade launcher on top of one of the humvees. fired the first shot.

They definitely were not happy about being woken up. And they definitely showed that. But then again, we did flash-bang them.

Like… ten times.

But they pushed back. Bullets flew past us like angry hornets. No doubt, they were all awake.

We were all cleared to hide behind our cars, disabling our dilatory and letting soldiers push footholds. We held our turf, returning fire. Our manned guns held back lots of pressure, but it spired to chaos when we lost our Gunnery man. After he was shot, another frontier soldier climbed inside the Humvee passenger side and pushed the body out. He took over and pushed the enemy line back.

The sound of bullet impacts was everywhere. We were parked with nothing but roadblocks and a fence for a bullet cover.

"They are flanking us to the left!"

Crap. They left through the front gate to hit our side. There was a meteor covering the left side slightly. It was the size of a small car, so it may just work.

I ran behind the meteor with three other frontiersmen following. I could hear the impacts of the bullets bouncing off the solid meteor. The next sound was the roar of motorcycles charging at us like the Cavalry of an army of old.

The crater of the meteor left a solid trench for us to work with. While I was crawling through the trench, the man holding most back from our side got hit. He fell down with a silent thud as the Sentinels started to charge.

What was I going to do?

Not the same mistake as before.

I poked out the top of my head and the barrel of my gun and shot.

And shot

And shot.

When my mag was empty, I popped,

Slipped.

And hit another set of rounds inside the M27.

Two mags left.

I continued to shoot, my spray of death hitting many and killing countless more.

My thoughts were darkening. My body was in fear. But in my mind, it was loud.

And angry.

Blood spewed, men fell. Bullets killed. I saw it all.

And I felt satisfaction running through my veins.

Pop. Slip. Hit.

One mag left.

I push myself out of my trench more and lie down, using the dirt and ground as cover. I fired at anything that moved that wasn't us. The recoil of the gun was making my shoulder numb at this point.

I'm scared.

click!

Pop. Slip. Hit.

"I NEED AMMO!" I shout as I fire off a few of my remaining 30 bullets. I can't look back to see if anyone even heard me. My focus was my survival.

"Jus' take this box!" Someone yells. A female.

A crate drops next to me. And it's filled to the brim with magazines. Some extended. All filled.

Five minutes pass, and I'm still holding back men.

Ten more minutes go by, and my arm is tired. I used almost half the box already. I finally look around quickly.

I was the only one holding the flank.

It wasn't until seventeen minutes passed before some other soldiers came back to help.

It didn't mean I had any kind of break.

Now, I was helping to hold the main assault on the north. I or my gun wasn't taking any kind of rest. The barrel started to glow red. Killing anything it hit.

Then, the gun burst into a small flame. But it continued to shoot.

Mag, after mag, after mag, I barely could see where I was shooting. The meltdown burned my fingers. My eyes. My soul.

The fire felt powerful, as if I was bending it to my will. Though, strangely. It burst more to the right of my chest. And it was getting hotter.

And hotter.

God my chest felt like it was on fire itself!

I finally stopped shooting and ran behind one of the Humvees. Over the yelling, I can still hear gunshots, though they sounded quieter as if my bullets numbed my ears, too. There was this... throbbing in my chest. Not on, in.

I grab my chest. It was warm and wet.

Sweat?

Until one of the prison spotlights hit us. It allowed me to get a good look at my hand.

That was burned… and bloody.

"I'm hit!" I yell without thinking. Dropping the burning M27. The flames stopped when it hit the dirt.

I could hear Dominic yelling.

"What time is it?!"

"12:45 Sir!" I could hear from an untraceable position Ijin's voice call out.

"Shit! Fifteen minutes still!" Dominic yells again. "Is anyone else hit?!" He yells over an explosion. I felt the dirt fling all over my body as I covered myself.

No one answered. Either they didn't get hit, or they died when they did.

"Private Rose! Fall back and treat the wounded soldier!"

"Yes, sir!" A female voice yelled over him, matching Ivys. Only this was a little lighter. I could hear heavy footsteps stomp over to me. She soon found me and stopped in front of me.

"Where is it?!" She yells over the gunshots.

And so I begin to wheeze, toppling over from my standing posture to sliding on the door and sitting on the ground. Clenching my teeth as I grab my peck.

"Okay, okay, it's okay. Stay with me, brother. Relax."

Ah. Brother.

My reminder of why I'm doing this.

Brother.

My brother.

He is somewhere amidst all this chaos, sick and weak. He needs me. I can't give up. Not now.

"How bad is it..?" I manage to spit from between my breaths.

"You're going to be okay, brother! You're okay!" She yells. I can tell she is trying to hide the alarm in her voice.

"The bullet. It's... it's inside..." I choke out. "I... I know how to fix this... but not with the bullet inside..."

I think back to when Coal got shot. He pulled the bullet out with rusty tweezers himself. Later on, at the tent, he told me that using your hands is the most dangerous alternative and that he personally would never reach that decision.

"Eight minutes!" Ijin yells.

"Hold our left!" Dominic yells. Now, with a fallen comrade, M16 — fighting with us.

"Drink this." Rose hands me a metal flask from her person. After I grab it, she rips my shirt and slowly lays me down. As she does, I bite the cap off the mysterious flask and chug the fluid inside.

The drink tasted like honey. The pain... it's fading... disappearing.

"It's bad," Rose says. "But it's not your heart. If we-"

Another distant explosion cut her off. The yelling soon came after.

"What was-"

"Nothing. Stay still. Here-"

Rose then hands me a Protein bar.

"Distract yourself with that," she commands, packing up her bag.

"Five minutes!!!" Ijin yells over bullets and death.

"Jus' sit tight, brother. We are falling back soon."

I don't respond. I just continue chewing the Protein bar. Looking down.

"COVER YOUR HEADS!" Someone in the distance yells.

Me and Rose both get on our stomachs and hold our hands over our heads. The most powerful explosion I hear hits. It sends dirt and rocks hitting everything and everyone.

"Get off your wound!" Rose yells, picking me up by my shoulder.

"One minute!!!"

"Start to fall back! Let's hope Trevor's body is among the rest of them!"

With bullets flying by us, flying above our heads and narrowly avoiding us, Rose begins to pick me up like a cartel baby.

"Where is your car?!"

"I'm a driver…" I say, weak.

"Allow me." I hear a voice deeper than hers.

Ivys?

Why is she here?

How is she here?

Where's Amaryllis?

"No… I'm too heavy…"

Rather, she didn't hear me, or she ignored me completely. I feel myself being handed off as I try to put pressure into my brain. Trying not to faint

"Where is… my gun…" I say a little louder.

"Forget it," she commands.

"No… please. Get it." I beg.

"Fucking hell, kid, you complicated!" She yells, frustrated. "Rose! Retrieve his gun and fall back!"

"Th' burned one?"

"Yes, the fucking burned one!" she yells.

"Yes, ma'am!" she says. I could hear the stomps of her boots running off.

My eyes started to flutter.

"Leave me…" I say faintly.

"Negative, I NEED MORPHINE!" I could hear the echoes of Ivys.

"Where's… Amar-"

The last thing I saw was the sky shroud around me, covering me in a blanket of darkness.

??? ???? ????

???

Somewhere in Texas

White. That's all that filled my vision. And it was like that for a solid minute, taking my darkness away, my comfort. All my comfort for the past forever was just hit with a blinding light of white. I thought I was dead. The pain in my body was gone when the light hit. I managed to feel my muscles twitch. My bones rest.

I jerked myself till I was sitting up. I was panting heavily. The room began to clear up for half a second. And for that half a second, I could have sworn I saw an angel.

"Will!"

"Amaryllis?" I groan, trying to ignore the pain.

"Lay down, Will. Your wound will reopen," she says in a shaken voice. She pushed my shoulder softly to help me lay down.

"Did… did we win? I mean like-"

"Yes. Yes, everyone is okay. Mhm," she nods. "Sit still. I'm going to go get the doctor," she says before letting go of my shoulder and walking out of the bright white room.

I begin to move parts of my body, yawning and moving my feet. I lift my arm only to see tubes in the back of my hand.

Ugh. I'm tethered with medicine.

I usually like to heal naturally. What I'm strapped to is a lot of plastic.

Beside the bed was the machinery of white and complex patterns of metal. It was something usually seen in hospitals as big as buildings.

Heart monitor, a sink, and the… fluid IV connected to me. It reminded me of the ICU back at home. Well, the school, if I could even call that home. With all the scarred, sad faces and bodies of everyone inside laying on the floor.

Ugh. My head is throbbing.

On my right, though, lying down in my bed next to me was my gun. The grip and barrel are covered in melted plastic and rainbow-melted steel.

So, I lift my left hand up. It was clean — but bandaged. I unwrapped it a little to see blood marks and burned skin. Luckily, I could still wiggle my fingers and move it well. No muscle damage or broken bones.

Still. That bitch really fucking hurt.

A couple of minutes later, the mayor from before walked in, with Dominic following behind. Amaryllis walked in almost too closely behind.

"Where is the doctor?" I ask.

The mayor chuckles. "I am the doctor." He had a husky tone that matched his robust appearance.

"Huh..?"

"Turned out he wasn't the mayor for no reason. His medical knowledge is outstandingly helpful. He could have basically healed half the citizens that are here."

"Ah... so we are back..." I let out. "Wait! Did we-"

"Yes, the mission was a success in small forms. We retrieved about 150-200 civilians. Many of which are doctors and some reserves for military branches and PMCs. Of course, with you playing a big role."

Oh. Huh?

"You held off our left flank flawlessly. Most of the charge was pushed back or-" Dominic cuts himself off.

"Yeah, don't butter me up." I sigh.

"Let's see here..." the mayor... or doctor pulls down my white blanket, revealing my bandages across my chest stained with sweat and a little blood.

He pulls a stethoscope from under the bed after inspecting me and puts the cold piece of metal on my chest.

"We are lucky you didn't get gutted in the lung. The bullet was stopped by your ribs. However, we are going to need the room back for another patient now that you're awake. Don't worry. The treatment I have given you will heal you quickly and well."

"How long have I been out?" I ask.

"Seven hours, twelve minutes, and thirty-two seconds!" Amaryllis yells before the doctor even speaks.

"Yes… see, this remarkable girl is correct. Speaking of such, she has something quite interesting she would like to share," the doctor says.

"Share?! How can we even trust her!" Ivys' voice came from the hall as she walked through the open door.

"Calm yourself, First Sgt," Dominic warns, stepping in front of her.

Honestly, I forgot he was in the room.

"Calm myself?! This girl says she lost her memory, yet she knows plants like she is Mother Nature with extensive medical and military knowledge?! Some of which she, or any non-military official, shouldn't even know about!"

As she talks, she slowly raises her voice, screaming the last part, causing Amaryllis to run around the bed and hide behind me. So, I sat up, though it was still aching.

"Yo. What are you all talking about? Someone explain," I demanded.

Dominic was about to speak until Ivys spoke first.

"Over the course of the raid, she acted as if she was my battle buddy. Quickly and accurately, she identified kilometer lengths with no scope or spyglass at hand. She named several united vehicles from where we were. She even yelled out, "Hostiles incoming north. Two clicks!" What person with serious dementia talks like that?!"

Dominic tries to calm her down. But my attention was back on Amaryllis.

All she does is stare back with her pretty eyes and I couldn't help but wonder who she even was. She appears out of nowhere and forgets everything important in her life. But when it comes to plants, medics, and the military. She is a top-tier polyglot for all three.

In all, I never questioned it mostly because everything she did with that information was to help us.

"Has she betrayed us?" I questioned. My tone firm.

"Excuse me?" Ivys says with almost a sneer of cold command.

"Has she betrayed us? Because all I see is her helping you and helping us." I say, or basically at this point shout.

"I hate butting in but the boy is correct."

The doctor is defending me. At least someone is.

"This has nothing to do with you, Adonis." Ivys retorts

"It's mayor Adonis to you," he says, holding his finger to his lips in a hush motion. "And in a brief, this has everything to do with me. We do need space for our new influx of civilians."

"Woah, woah hold on…" I put my feet down onto the cold floor — stars rushing to and through my head.

"Easy Will," Amaryllis says softly, grabbing my arm.

"What are you complaining about?" I ask Adonis.

Dominic answers instead. "On the way back, she overheard me with commander Zu talk about where these civilians could go. And… I guess I should let Amaryllis explain what she said to us that the mayor is quite excited about."

I look back up at Amaryllis. She was still looking at me, worried.

"Amaryllis?"

"Yes. It's true." She sighs. "A company working for ECTO named Iron Pulse Industries has a base in Houston made for nuclear fallout protocols that could theoretically sustain more than three thousand people with advanced food cropping systems. Located at 32.64145° N, 97.13888° W. Requires a security access of five and above."

"And that fact is correct. That location she mentioned is Fannin street, in Arlington" Alder says. "There is a launch site there owned by ECTO. As stated by her, Iron Pulse holds a bond with ECTO as a supply for advanced war materials.

"Is Iron Pulse a branch?"

"No. Just another PMC, although it was the most well-funded and dangerous PMC in the world. I'm surprised we don't see them on the road. But then again, we don't see the United States Army either. I pause and look down at the floor.

Contemplating.

Then, a nurse runs inside. Panicked like a freak. When she does, I notice her face. Besides her over-sweat, her dark hair covers half of it. However, it didn't do a good job of hiding the scar that formed over her face. From her cheek to her forehead, a double-sided axe shape was scarred into her face.

Burned.

"He's awake!" she shouted.

"Beast?!" I almost screamed.

"No… not Beast," Amaryllis says, grabbing my hand… from the non-burned side.

"A boy. An extraordinary boy. One that kinda looks like you."

My heart skips a beat as I pull out the tubes in my arms and stand up. Dizzy from the blood rush.

"Bring me now."

September 13th, 2035

8:09 AM

Corsicana, Texas.

Could this be Jack?

I'm standing outside a steel-framed room. Windows framed and bolted to the walls. It's almost cage-like.

Did he do something? My fingers twitch on the butt of my gun — which I'm using as a cane

Could this be him?

The one reason I am here?

The steel door opens a crack, and the nurse comes out, closing the door behind her and taking off her medical mask.

"He seems to have similar properties to you. The DNA test also came back positive."

"Then let me see him?!" I let out, moving forward until Dominic grabs my shoulder.

"I'm sorry. But the way he has been contained tells us only that whatever sickness he has could be deadly and contagious— putting risk to the rest of life we have so far.

"Aside from that, the boy requests that no one sees him," says the nurse.

"But I'm his brother!"

"Listen… Mr., Light-"

"No! I have to see him!" I shrug off Dominic's hand before he grabs my arm. For a split second, I thought about pushing him away. And what colour I wanted my casket to be.

"He specifically requested that he would rather have anyone else in the world see him than his brother."

What?

"He said, and I quote, "Do not ever bring him close to me. Ever." I'm sorry, but it's true."

Woah.

Ouch.

Dominic begins to let go of me before Amaryllis starts to hug me.

"It's okay. I'm sure he is trying to keep you safe."

"By allowing anyone else to see him but me?" I say, hiding my shout. Yes, she could be right. But for hell, he would rather let Alic-

Oh, wait. My mother would never let me call him either.

Was that by request from him, too..?

I swallow the rest of my words quickly, holding back claims and tears.

"It's fine."

"Will?" Amaryllis says with a worried tone.

"I'm okay," I lie. Not just to her but myself.

He never wanted anything to do with me. Guess the best thing to do is to not think about it. Distractions are dangerous, especially in the combat situation I'm in now, aka life.

"When will we go to see the new shelter?" I ask, hoping to change the subject.

"Three days. We need everyone to be ready."

"Everyone?"

"Everyone. The whole camp. All the soldiers and civilians within. The mayor is really thrilled with what Amaryllis said. Too excited even."

"Alright." I nod, walking to the very rusty elevator shaft. When I make it close enough to hold onto the wall I pick up my crutch of a gun and rest it over my shoulder. The melted plastic and rainbow iron really needed to be fixed if this thing was ever going to fire again.

This medicine —whatever he gave me heals so quickly. Back then, my father would tell me it could take months of rehabilitation before a wound was ready for action. Now it takes at max days. Time changes a lot.

I grab the button hanging on a rope and press it. A loud buzz filled the air as the elevator cranks its way up to me.

"Wait! Will."

I didn't turn around to Amaryllis's voice. I was too annoyed.

"Give him time." I hear silently behind me as the elevator makes it to my level. Adonis. "Besides, I need your help. We need lots of information on the bunker."

Yes. Please leave me alone.

I step inside, not turning or saying anything as it goes down. About a minute and a half goes by before I reach the bottom of the building. When I stepped off the elevator, there were these two Frontiersmen; they walked up to me. They only added confusion to my fuming head.

"This? This is the young grim reaper we heard about? What happened Ironsights? Feeling down?"

"Huh?"

"You Ironsights. You are apparently the Irons we have been hearing all about. Looks like the thumb sucker can be of use. So why do you seem down?"

"Ts' nothing."

Unexpectedly, he kicks me in the gut, and I stumble to the ground.

"Good. Because I want your gear."

He tries to kick me again, but this time, I grab his foot and twist it back, causing him to fall on my side perfectly. Though the second soldier grabs my leg, I manage to kick him in the face and return to the wrestling motormouth. He gets below me, and I smash my elbow into his nose, hearing a lovely crunch that sounds like music.

He yells and kicks himself off of me. The blood from his nose covers his stupid moustache, draining down his cheeks as he yells slurs, rolling around as I stand up.

"I'LL FUCK YOU UP!" The other soldier yells before charging at me. He throws a wild haymaker, and I dive it, slamming my palm into his throat. He chokes and falls on his knees. I step up, standing above him and looking down on him. He continues to try to catch his breath. My vision was blurry. But I distinctly remember grabbing his head and slamming him against my knee. Knocking him out as he falls on the dirt filled floor, face first.

"Impressive Irons."

That recognizable voice enraged me more as I turned and saw that instead of calling me shrimp, Monroe called me this Irons character. He walks up with his little crew of two. They were the Hispanic and black man I saw at the school.

"Heartlessness like that makes you survive. No one lives without being ruthless. This is what earned you thirty-five bodies."

Thirty-five.

At that moment, I realized what he meant. I was about to tear up for ruining those lives.

They were going to kill us if I didn't kill them.

He was right. Will had to die for Ironsights to live. I'll make him my next body. I would do that no matter if it mentally tolls me or physically kills me.

September 14th, 2035

9:00 PM

Corsicana, Texas.

Nothing today. Just a bunch of commands and whatever else.

September 15th, 2035

8:03 PM

Corsicana, Texas.

    Nothing today.

September 16th, 2035

9:00 PM

Corsicana, Texas.

Nothing today, still.

September 17th, 2035

9:12 PM

IH-45, Texas.

Nothing today. Except that, we started the trip. We are on the road — again.

September 18th, 2035

11:03 PM

IH-45, Texas.

Nothing today. Just that I... killed some Sentinels, who were trying to stop us.

September 19th, 2035

11:22 PM

IH-45, Texas.

Nothing today.

September 20th, 2035

11:41 PM

IH-45, Texas.

Nothing. Nothing at all.

September 21st, 2035

11:53 PM

IH-45, Texas.

I killed more Sentinels today. Total so far? 96 kills.

September 23rd, 2035

12:23 AM

IH-45, Texas.

Nothing. More kills. Nothing… else.

September 23rd, 2035

11:57 PM

IH-45, Texas.

Halfway there.

September 24th, 2035

14:16

IH-45, Texas.

Hello. My name is Amaryllis. And I am the new caretaker of this journal.

After reading this journal, I learned so much about Will that I didn't know. Like how he was a normal kid before a "rain of fire." I didn't even know the world ended before I saw him write it inside here. Meteors attacked us, but we survived. And as he writes, *If we can see light,* **There** *is hope.*

Will threw this journal outside a fire pit when we had our second campsite stop. Most likely, he was trying to burn it.

Also, He doesn't go by "Will" much anymore. If anything at all. I'm really the only one who calls him that. But all the others call him "Ironnsight." Or "Irons" for short.

He's changed.

When I last saw him, he was a friend. The one I trusted.

But now... that light in his eyes... it's gone. He barely speaks and doesn't smile. He is always with a cautious look on his face. The Will I knew was hidden behind this... *machine*. He has become a soldier who has nothing to mind. Although he is respected by other soldiers, he is mostly alone. It is almost as if he avoids everyone. I overheard a couple of soldiers spreading the role of titling him "the God of revenge."

That anything he ever aimed at is dead.

That he was inhuman.

That his one hundred-plus confirmed kills are his title to a true monster of a soldier.

It's been only seven days since we left, and he killed that many people. He almost started to scare me until I realized I was the only one who knew him. *I* was the only one who could help him.

Sadly, I have been busy, too. The mayor has been asking every question possible regarding the base. I had to keep reminding him that I have dementia and that I don't even know who I am. He also had the Nurses prepare me to meet Will's brother, Jack.

The doctors have been "training and preparing" me for this meeting. The doctor insists it's very important

I'm prepared.

Overall, I have been with the soldiers more than the civilians. The plan that Adonis and Dominic are following is that when we capture this bunker, we will head back to the school base and pick up the rest of the civilians.

It was dinner time. What surprised me was that Will himself brought me an MRE. He knows I ate one already so I knew this one was his, and his last one. He even *cooked* it for me.

"Thank you, Will."

He just nods and walks off. The food is really good but it would be better in his presence **even** if he won't talk to me.

So, I get away from the civilian feeding ground and follow his path. He usually eats alone, standing on guard on some post. He always takes a lookout job that opens.

After a little I see him sitting by a leafless tree. I walked up to him and then tapped his shoulder, causing him to flip around me and aim his gun at me. Dropping his food and gathering his thoughts, he lowered his rifle.

"Sorry"

"No no… I'm sorry, your food—"

"Tis' fine. Wasn't hungry anyways."

"Please take mine," I beg. "It's really good."

"No thanks… you take it."

"If you don't eat it, I'll throw it away!" I say firm.

He takes a long, long sigh before grabbing the MRE from my hand. "If I eat a little, will you eat the rest?"

"Mhm," I nod.

He takes a bite before handing it back right away.

"Eat."

"Please take more."

"A deal is done." He hands it back and turns around.

Hmph. Touché then.

I learned that word from reading.

I sat next to him while he was standing and leaned on his legs.

"What are you doing?"

"Eating."

"You're on my legs."

"I want to be here."

"I'm on duty."

"I don't care. I'm staying." I complain.

"Fine."

He didn't fight it or move while I leaned on his legs. I finished eating and leaned deeper under the black blanket of the night sky. After an hour, twelve minutes, and four seconds, he shakes me with his foot. Practically wakes me up.

"Come on, it's late. You need to sleep."

"Come with me?" I ask, looking up.

"No. I'm working," he says, not even looking my way.

So, I give up.

"Okay," I sigh, getting up. As I walk back to the Humvee I am hoping he will stop me, to ask me if I want to pick flowers with him.

But he didn't.

I want to cry.

So, I did. For twelve minutes and two seconds before, I ran out of tears. He is somewhere in this… cocoon. Hidden.

What changed? What changed *him*?

September 25th.

08:16

The next morning was us packing up to go on the move again. When I saw Will, he was loading a T500 War flare into one of the trucks with three other soldiers. So, I tried walking up to him.

"Will!" I called out.

"Hey kid, Relax. Irons is busy," says a soldier walking in front of me, blocking me off. He was an Asian soldier, tall and strong. I assume this is the Monroe I read about. The best and worst part about him was that he didn't scare me.

"Will!" I yell, calling over Monroe. "Can I help?"

He turns and looks at me, his face dirty with soot. Yet his tone was as soft and as simple as ever.

"Go help Ivys pack up my car," he says, turning back to strapping the War flare.

"You heard Irons." Monroe snorted, not moving, as he crossed his big arms.

"His name is *Will*," I say. Well… at this point, I practically yell.

"His *new* name is *Ironsights*. In which he *earned*."

"Monroe," Will said. Softer.

"Yeah?" he spins around.

"Be nice. Or I'll shoot you dead right here," he warns, not even turning to face us.

That was *not* Will.

Not at all.

"Will, it's okay. I was leaving anyway. Make sure you eat," I tell him, turning around and walking back towards the Humvee.

I see the pretty woman loading up .300 sniper ammunition into her sniper rifle, and some other Guns on

171

the table in front of her were all empty of Magazines. One I noticed was an oddly shaped M27. ***Modified***. Modifications I have never seen.

"What are these silver parts on the railing?"

"Door hinges," she answers.

"What? Door hinges? *Why?*"

When his gun had the meltdown, the gas chamber burst and was left inoperable. Although he easily could have gotten a completely new service weapon, he insisted on repairing the one he has now. Welding steel door hinges to the open metal.

"But what about *this?*" I say with slight astonishment as a hand-painted *wolf* was shown running across the stock of the gun.

"Exactly what you see. A wolf that he painted himself."

"The magazine has a beartrap painted on it," I point out.

"It's his gun. I have no idea what the meaning of it is."

"Wait I think I-"

Suddenly, my head started throbbing with pain as I tried to remember the last time I saw a wolf.

"Sorry. It's hard to remember," I say gripping my forehead that seemingly was about to explode.

"You're fine. Drink some water," she says, putting down the gun.

"Amaryllis?"

"Yes?" I answer as the pain disperses.

"I want to apologize."

"Excuse me…?" I say, almost wondering if she is going to hit me.

"I accused you falsely. It was not right to accuse you of faking loosing your memory. Observing you, I can see that even if you did or didn't, you still care about Will."

I nod once. "But… he has changed so much these last few days."

"I know. I talked to the Colonel about it. The response was short from him, saying he *"couldn't do anything about it."* They did talk, but he wouldn't say anything… *meaningful.*

So, I was silent for a moment. My worried side was clearly shown because she started talking in a softer tone. More relaxed this time.

"He is just lost. However, he does have a soft spot for you. So, I need you to do one thing for me.

"Yes?"

"Stay with him. Please. He has us, but he only sees you."

"Yes, ma'am."

"There you are!"

I practically jumped as I whipped my head around. Adonis was speed-walking in our direction.

Then I remembered why.

"Oh! I am so sorry! I completely forgot about my meetup with Jack!"

"How long did you sleep last night?" he asks.

"Five hours and Twenty-Two minutes exactly.

He paused. Finally, cutting silence with "Alright. Come along."

And he spins around, walking off.

"Okay!" I say, running behind. "Bye, Ivys."

She waves before grabbing another mag to load into another gun.

The mayor seemed to forget about me being late. When we approached the Boys tent, he started reminding me about everything that he told me before.

Don't get too close, don't touch him, be gentle with him. A long list of commands.

I remembered them all.

"Today, you will just meet him. You have an hour to bond with him before we get back on our path.

"I'll do my best, Sir." I say, Nodding.

"He's inside," he says, ending the brief and walking off.

And so, I stand, looking up at the yellow tent as if it's a cave of mystery.

I take a deep breath.

And walk inside.

September 25th, 2035

08:45

IH-45, Texas.

Flowers are beautiful but I never expected to see some so soon.

Walking inside, I instantly recognize the pedals of color blooming all inside on the floor. When the doctor told me his room would be filled with "Flora," I expected inside pots because the doctor also told me that they were safe to touch with gloves on.

Though not from the boy.

Almost like a younger, mirrored version of Will.

His hair was longer, but they both had the same messy blond-silver mix hair. But he had... plants. Vines came out of his scalp, wrapping around his hair a little. He also had the same flowers sprawling out from under his shirt, through his arm sleeves, and around his neck. All he did was look to the floor from him sitting on his bed covered in colorful flowers. Each one finding its way connected to him.

"Hello, Jack. I'm Doctor Amaryllis. I'll be taking care of you." I say, keeping myself from staring too hard.

I may have lost my memory. But I have no idea if those flowers are normal at *all*. He nods and extends his hand, waving his flowered stricken palm.

"When they said you were special. I must admit, I was not expecting this."

"This?! As in the monster I am?!" he suddenly burst, spreading his arms out.

"In my experience..." I started calmly.

"...anyone who can grow plants from their body is a superhero rather than a monst-"

"In your experience?! I was told you lost your memory, so what experience?! If anything should come to your memory, it's that *this is not normal!*"

And finally, after that burst, he crawls into a ball and begins to sob. The flowers around him — around us started to close. The scene was unfolding as if it was a cold winter day. All of them slowly closing like a wave of air. So, I took the hint.

"So why do you think you're... that?"

"Just look at me. I'm a freak," he says from behind his knees. "I'm not *normal.*"

"So am I," I say. Trying to look at him.

"Ts' different..."

"Only because you say it is." I begin, moving towards him and sitting next to him in his flower... bed.

"Your power is unique. Awe inspiring even. But past that, you're still a normal kid. Beyond everything that's happening, beyond this unique special power you have, you're still *Jack.*

I wanted to forget Adonis's' first term of not touching him just so I can hug him, but I my composure. Such a tiny thing fighting between two wars right now. One with us, and one with himself.

"I wish I wasn't sick. I wish I were normal. Like really normal," he says in between sniffles. But the flowers around him. They started to open slightly. I can't give up now.

"Right now. You are the most unique, special, and resourceful person to everyone here."

He doesn't say anything, but the flowers begin to open up more.

"That little ability of yours can save the world. Everyone's world."

"What?" he says, lifting his head. "But how?"

"You can grow *flowers*, Jack. The earth alone is having a hard time growing *grass* after the rain of fire. I can't say it's not doing anything, though. Some trees are alive. Because if they weren't, we would be... um... gone."

He looks down. "What does that have to do with me?"

"If we see your reaction with real — actual living plants, we could try growing the world back slowly. Allowing life to flourish," I explain.

"You are a valuable seed."

And a second later, he hugs me unexpectedly, making all the flowers in the room bloom.

It was magical.

"I promise I won't infect you."

I couldn't pry myself any longer. I wrap my arms into an embrace of safety and take his head into my shoulder. I could feel the roots growing around my arms and my legs, tangling around me.

So, I slowly pull away as he lifts his head. We both were now inspecting my newly tied body of vines and flowers.

"Um… does it hurt if I cut these?" I ask, lifting my arms as wings of roots and color hang from my arms as if it was an old grass-woven coat.

He laughs a little, seeing me. "You look like a bird," he teases.

"I do?"

"Mhm, and don't worry. The pain depends on how long the vines are. From where you are, it will be just like feeling a pinch for me.

"I still don't want to hurt you…"

"Trust me. This is not as bad as it can get."

Ouch.

"Okay. But I don't have scissors. Maybe I can-"

"-I do!" he shouts before pulling a pair of small mayo scissors from his hair.

"Wha- how did you get these?" I ask as he hands me them. They are small and sleek. So that gave me a little of an answer before he did.

"Pickpocketed the other doctor. You can't blame me… I kinda do need them more than him. Besides, I trust you a little more. You seem… genuine."

I guess he can give off the vibe of "being too untrusting" when you see him working. He's very clumsy when it comes to leadership. But everyone just executively voted him as the mayor of the last base because he does know how to heal and help anything and anyone wounded."

"He's really trying to help too, you know…" I say as I snip one of the vines as quickly as I can. I see him flinch a little, but he doesn't show signs of pain… yet, at least.

"Did that hurt?" I ask. Okay, maybe with a hint of panic in my tone.

"No. Just… just a pinch," he says, slightly smiling. "Can you crawl out of the ones on your legs?"

I don't answer. Judging by that, they must have hurt more than he said they did. So, I begin to slide through the ones on my thighs. They were a little tight, but I got out of them eventually.

"You probably shouldn't tell the doctor we made physical contact yet. You can guess how he will react."

"I might be young. But I ain't stupid," he says. Smiling again.

And over the course of the next half hour, I get to know him as he gets to know me. Or at least what I remember of me. He is currently eight, and his birthday is in forty-three days, sixteen hours, and thirty-one minutes.

Also known as July 4th.

Similar to that, he loves history, at least anything involving the past two thousand years. He can't go back farther with knowledge. He's not really into dinosaurs and evolution.

I do wonder if we are evolving right now.

One major thing about Jack is his trust. He's very hard to gain trust with, which nowadays, is very understandable. He found trust in me, though. Why? In his words, "You're familiar. At least I feel a familiar force between us. I don't know why, I never even met you before today."

"That's strange to say," I responded. Though In reality, I get what he means. Some people I look at I get that feeling. The citizens mostly. I don't know what it could be. Past memories? If that was the case then Jack would remember me.

I do only mostly wonder who I am, more than anything.

Maybe he just sees Will in me through my actions. Well even if that was the case, I know that he doesn't like anyone even mentioning the topic on Will.

So I wont, especially on first interaction.

I do also learn interests and dislikes of his. The conversation after that ended shortly after. Thirty minutes was all that I was given. And it's not the best amount of time.

"Amaryllis?"

"Yes Jack?" I respond as I'm beginning to stretch for my exit.

"Please come by more."

"Of course. I will try." I say, patting his head, before getting up and walking out.

The area outside really cleared up from when I was last out there. Doctor Adonis was walking towards the tent with a nurse following close behind, guiding a truck. I'm guessing that's Jack's ride. Strong and secure.

Speaking of which, afterwards I went to the Reddened Humvee that belonged to Will and sat in my usual seat. He wasn't there. No one was.

But there was one thing that caught my eye.

When I look forward, I notice that on the rear-view mirror, a glister of gold was hanging. The style of it represented a clear bullet casing of solid soil. What was weird about this… pill was that it had a *seed* inside.

I mean sure, that could be normal. If it wasn't that a beautiful green plant sprouted from the inside. And it's *growing!* Even with no water *or* sun!

I was going to grab it to inspect it closer until the driver door opens, and Will stand there looking at me scramble to sit back.

"You doin' alright?" he asks.

"Mhm! I'm okay!" I answered quickly.

Two minutes later and everyone is back inside. Monroe was next to Will, while Dominic and Ivys were next to me. And I was moved to the middle.

"Are you sure you're okay to drive?" Ivys asks Will.

"Yeah, I can breathe, see, and hear. I'm alright with driving," Will answered.

"Alright." Ivys closes off, looking at Dominic. They seemed to exchange glances of words, because all the Colonel did was raise his hand and lower his eyes.

"Follow Zu," Dominic commands.

"Yes sir."

And so, after the roar of the engine, our path continued down the foggy road. And down my foggy mind, where the only thing was going on was Will.

September 25th, 2035

21:11

IH-45, Texas.

Soldiers lost their lives today.

But not Will. In fact, I think he may have saved us.

9:22 AM

Over the course of the quiet ride, I stared at the pendant that hung from the rear-view mirror swinging around from our rough terrain of road.

The car was filled with chatter on what the meteors could be made of. Some emitted plasma, while others were just large and purplish in color.

"The hell does plasma even do?" Monroe asks. But before I could answer, the response came.

"Usually, it's pretty neutral, but it can rather be used as a battery or a battery killer. And normally, it's pretty neutral because it's not radioactive. However, full strikes of plasma can be deadly."

"You seem to know a lot about them," Ivys said. "When did you become a thermonuclear astrophysicist?"

"My brother worked for NASA as the owner of the ECTO branch; I picked up tabs of little things from him."

"What was his na-"

"Road's blocked." Will cut into Ivys' question as he stopped the Humvee.

But he was right. The whole highway ahead was cut in half by a giant black plane acting as a barrier for us to pass, the right wing was torn off, and the left wing faced the other side of the highway so it could be as well. Though on the side of the body in white spray paint, read:

TURN AROUND!

"What kind of plane *is* that?" Ivys asks.

"Must be an airbus. Those reach up to 200 plus feet long. Gotta cut around it," Monroe answered.

"Yeah, but why is it *black?*"

"Racist," Monroe chuckled.

"Says the *Pilot,*" Ivys sighs. "Now go-"

KABOOM!!!

"The Fuck was…"

Weapons live!!! Dominic yells, throwing the door open and stepping out of the Humvee.

Then it finally reaches me.

Gunshots, impacts, explosions.

The air was back to its fire-filled smell. The sound of the outside and the camp followed us. And it finally reached us. The yelling and the gunshots that followed scared me once again. And this time, I was trapped in it.

Will was grabbing his gun and inspecting it. Checking safety and ammo.

"Stay here," he commands.

"Mhm" is the only thing I can say without going against him.

And he exits, running out and behind the Humvee where the rest of the troops are running while I was scared inside of that Humvee. As I covered my ears, I saw Will, who was charging right at the sound of death.

Not walking. *Charging* towards his voice.

My astonishment passed quickly because I realized…

Jack.

What if he is in trouble? He is back there.

I have to at least check. Will went out there with no pauses. No second thought.

I have to be like him.

So, I calm my breath. Before opening the door and running. Running behind trucks, tanks, artillery…

The sound was severely louder outside than what it was inside the Humvee. So, I continue running, covering my ears, until I see Will. The soldiers next to him hid in the rubble of dirt, covered in it. Yet Will... he was there holding the line, holding an assault —laying down and firing like a turret.

His eyes — they were just as they were when we were driving. Nor did his face structure change while we were in the Humvee. The only difference between him was that his neck and arms were flexed. Veins spiraled around them as the gun continued its loud cries of death. His face was... soulless. He looked as if he was doing boring homework.

I force myself to run past him. He didn't even turn from his fight. He was locked on the enemy. He was, in that moment, the perfect soldier they imagined him to be. Poor Will.

Finally, I spotted the truck carrying Jack. The Kraz-6322 was sitting silently, rumbling as the firefight continued. And I knew it was Jack because Alder and his Nurse were inside.

"Get the turret and the Grenade launcher over here now!" I could hear Zu yell in the distant gunshots. I don't see her, but her booming voice caused two Humvees to drive out of line and onto the right side of the brigade. I ran up to the truck and knocked on the window. Adonis lowered the window and in a very loud voice and tone, said, "What the *hell,* Amaryllis?! Why are you out here?! You could get killed!!

"I'm just... checking on-"

"Jack is fine! Just get inside quickly! It's dangerous out there!" he says, raising the window and flicking a switch.

Suddenly, the truck jerked as the heavy plated door opened on the side. The greenery was immediately visible when I made it to the entrance.

"Amaryllis?!" Jack called when I saw him as if he saw a ghost.

"Hello Jack... are you-"

"Behind you!"

I spun around to see a well-kitted man. Strapped neck to toe in black body gear, he had an ugly sneer crossed on his tatted-up face as he carried an AK-47 in his hand. Raising it up, he shot six times at the roof of the truck.

Warning shots. He was not with us.

"Get over 'ere boy," he commands.

"No! I won't let you!" I say, backing up to Jack, protecting his body behind mine. I knew Jack was scared because I was too. He was large —6'3-ish with well over 200 pounds on him, not including his gear. My protection was as worthless as meat armor.

All of a sudden, I heard a gunshot.

One that didn't hit any walls or fly in the distance.

It was smaller. More powerful.

I immediately checked my chest and body for wound marks; then I looked at Jack, who was fine, too.

But then I looked at the man. He now wore a wild expression on his face while his head tilted slightly.

"Is he…" Jack starts before stopping. We both watched as blood trickled around his shoulders, down from the side of his neck. He slumps down on one knee and flops over on his arm. Behind him, Will lowered a Sig M18 with its barrel smoking slightly.

Will. With no pain on his face, his lip twitched slightly as he stared inside the truck, not at the plants around or at me. But straight at Jack.

September 25th, 2035

09:44

IH-45, Texas.

Will, as soon as he sees all of the flowers, looks away… while grabbing the body of the man and pulling it out of the truck. The only hint of expression he had was his eyebrows furrowed up as if he was worried or confused. But he keeps a straight, silent face as he slams the door shut. I can hear the airlock after a few seconds.

"Who was that?" Jack asks from behind me. Now, his face looked like he saw a ghost. And I understood why.

"You..?" I begin before stopping myself. I can't tell him.

"That's just one of the soldiers who had saved you before."

From the Sentinels?"

"Correct."

"I want to meet him. 'Jus to thank him."

"He is a very busy man. Though I will try to tell him." I promised, hoping I could keep it.

"He reminds me of my dad…"

"Excuse me?" I ask.

"My dad. He was a real soldier. But barely anyone could have guessed by looking at him. He was friendly. Super friendly. But he died on the battlefield years back, in actual warfare. The service treated him well… the field didn't. That guy reminded me of him," says Jack as he sits down on the flora-filled floor.

"I'll see what I can do," I promise.

I showed myself calm. But I was secretly panicking while talking about Will. It was mostly on the fact that I didn't wanna ruin his mood more, especially with what happened.

"Are you okay with what happened, with what you just saw?"

"It's grown on me. The people who took me from the hospital did way worse in front of me."

All that got from me was a sigh.

"I'm okay, really."

"I don't know whether to be glad or not," I say. I tried avoiding admitting how much that scared me for him. No matter what, this boy, or anyone for that matter, shouldn't be exposed to this many sorrows of death to the point of taking it as a natural cause. Even though it is, it's still sad to see how the environment itself changed us… and Jack.

Silently, I sat beside him. We could hear the faint yelling and gunshots outside as they hit the truck or anything around. Both of which were loud and powerful.

The truck did its job, though. Defense and mobility were key factors in designing the Kraz-6322, so its sole purpose was to protect anything inside.

Which was us.

"We're gonna be okay, right?"

"Of course. The metal on this is mixed with titanium and steel. Penetrating this mixture will take a 950. JDJ. Even with the speed of 1500 meters…"

He looked at me as if he didn't understand. I'm guessing it's because he doesn't.

Kinda obvious on my part.

"To make my words simple, we will be safe. Wi… That *soldier* saved us while ending our only problem." I say. Catching Wills Name in my throat.

"I know."

After that, we just started off our talk from a couple of hours ago that we didn't finish. Our talks led to food, places, and future plans. All of the talking came to a halt after about 30 minutes when Jack started informing me about what happened in his story before the rain of fire. The doctor who ran his year-long "Operation" was nowhere in his vision of anybody we met; he assumes that she was killed during or after the event. *Forever ago*, in his words.

The doctor would never get physically close to Jack. He told me he had nightmares about her, though. She would wear a full-face mask with a foggy shield whenever she was running… tests on him.

He was put through a lab like he was a mouse. That doctor would make him run on a treadmill for numerous hours at a time. She would run multiple fluids in his body that were built to "Help" him with his ability.

He named them all for himself.

One that he called "the metal serum" was pumped inside one of the vines. He described the feeling afterward as relating to having a throbbing headache everywhere on his body. Stabbing each of his organs. He called it the metal serum because it looked silver inside the syringe.

There was "The Ghost Serum."

It knocked him into a nightmare that lasted a week in his mind, but an hour in real-time.

"The fire serum" was the one he feared the most.

It made him physically unable to move. And he described it as if they were lighting all the vines on fire... slowly burning each leaf off for the next nine hours of his life.

"Of course, the world ending is horrible. But in a way, it was my saving grace. I experienced more in a day out here than in my white cell. Though the sentinels are still scary."

"Understandably," I say.

"Amaryllis?"

The voice that seemingly came from all around actually came from an intercom from the roof.

"Doctor?" I say, raising my voice so wherever the microphone was, it could hear me.

"We were successful in pushing ourselves past the ambush and made it to Madisonville. Though some soldiers held back to defend our line. When we find a place to settle, I'll be there to let you both out."

"Okay," I said before hearing a click, and the buzzing of the roof went silent. To be honest, I had been so worried for Jack that I didn't even notice we moved. And that worry worsened when I saw the plants around him turning a lighter green. Almost yellowish. I pretty quickly figured out why.

"Don't worry, Jack. When we stop, I'll grab you something to eat," I say, my own stomach rumbling at that thought.

"Am I that obvious?" he asks sheepishly.

"Plants lose color down the stem when they lose too much sunlight... or, in both cases, food. It's natural, and it's just telling anything around that it needs sunlight to photosynthesize. In your case, you're the plant."

"You're impressively smart for your memory loss. How did you learn so much in such little time with... not even a single plant around?" he asks eagerly.

So I tell the truth. "I didn't forget everything, Jack."

"Huh? But…"

"I don't know myself… but I'm guessing what I do remember were key things in my last life. But they are all so unique I have no idea what I could have been in my last life."

"One keyword I can compare you to. *Amazing.*"

"I hope so. But I do have a lingering thought that I was a bad person."

The truck suddenly jerks to a halt, and after about a minute, the door flies open. Adonis and the nurse step inside, wearing gloves and all. Of course, he chews me out for not wearing protective gear. However, after his constant reminders, he finally realized I was perfectly okay and wasn't a tree or something. So, he let out a long sigh and dismissed me to go get food for Jack.

In the civilian area, they were unloading everyone and passing out sleeping bags and food. It wasn't the best. It was… Broth and cornbread. But it is food, and it's enough. I get it both back to the truck that was now a makeshift tent, with the iron door from before being replaced with a blue zipper door.

When I walk back and knock on the wall of the truck, the nurse from before steps out.

I really need to learn her name.

She had no gloves on, though she still had a mask on.

Guess Adonis' protocols changed. They were getting comfortable, which was good news.

Yet, of course, my mind kicked in a reminder of one person.

She thanked me and took the food. I did hope he could eat it.

When Adonis walks out of the driver's side of the truck, I walk up to him.

"Doctor, I request to be dismissed. I can't parse out why now, but I will soon."

"Dismissed." He says, sinking back into his clipboard.

"Where is Will?"

"Still holding them back," Adonis answers as he scribbles something.

The doctor comes out and calls for me. "Amaryllis. Come in. Poor boy has not been eating."

"I'm sorry. I would love to, but I need to see one of the commanders," I say. Before turning around, I managed to shout out for Jack to be safe. There were the camped soldiers on the left side of the civilian camp. No tents, no chow table, just a jumble of jeeps and tanks with soldiers all around eating from their own packs.

And, of course, I saw Will's Humvee. But no Will anywhere near it. Only a serious-looking Ivys holding a scratched-up sniper on her shoulder.

"Ivys!"

"Amaryllis?" She shouts, turning.

I run to her and grab her arm softly. "Will?"

"Fighting them damn mutts. We almost got surrounded."

"What?! How did Will escape?!"

"*We* escaped by drilling right through the plane with one of the tanks. While the ground troops defended the new escape in the plane, Will held the sidelines all by himself.

"*Is he alone?!*" I practically scream. Panicking.

"No. Soldiers are now with him. However, yes, he was alone from where we moved."

"He was *surrounded!* Why didn't you stay with him? He's not immortal!"

"I tried explaining that to him," she sighs. "I said that the whole front side is surrounded. You wanna know what he said back?"

God, no, I didn't. I shudder as if I'm hearing about a cursed ghost that possessed my body.

"What did he say?"

"*That's fine. I won't have to aim then,*" she mocks in his monotone voice.

"That *Idiot!*" I finally screamed. Causing heads to turn.

But I didn't care.

"Relax. Colonel Dominic is taking command over there. They will be back before nightfall."

I looked up at the sun, and it was about 0200 hours. I sigh to myself as Ivys walks past me to the driver's side of the Humvee.

"You took his car..?"

"He told me to," she says as she opens the door and steps inside.

"You're going back," I answer for her and myself.

"I'll be fine, Amaryllis."

"Take me with you!" I burst out, grabbing the outside door handle.

"Hell no! It's way too dangerous!" She says, slamming the door and basically pulling me with it.

"You need accurate readings of distance if you're sniping!" I yell before she gets the next idea to roll the window up. "I can help with that! We won't be too close to the danger with your vantage point!"

"What do you think Will would want!" she yells. "Stay here! That's an order!"

As if there was an invisible switch that was pulled, my body froze. I wanted to argue, but I couldn't even *speak*. This... feeling was impossible to pull away from.

So, I speak as if I'm speaking to myself, going around this invisible barrier.

"Will is *trapped* in that *lifeless, psychopathic* state. If he continues to live in it, he will get himself *killed*. You have given me the order to get him out of it, which *I promised* to follow. And you *dare* make me defy my orders? My duty lies in him and this army *commander*." I finished.

She leans over and pushes the temples of her eyes with her fingertips.

"Please," I pleaded one last time softly.

"Just get in," she breaks.

I won.

I silently go to the passenger side and make my way inside.

"You don't have a uniform. Under the seat, there is a shirt. Give it to me," she commands, beginning to drive past the buildings around. As she drives, she crashes into a small bike that is in the way.

I reach down and grab a maroon T-shirt. It was dirty, and it had giant dark blotches of red on it. It looked as if too much paint was spilled all over it.

She takes it from my hand. Holding the steering wheel with one hand, she bites it and rips a long strip down the middle of the shirt. She pulls off this long strip of cloth before throwing the rest of the shirt out of the window.

"Tie this around your thigh," she says, throwing it at my lap.

Though when she does, she pauses for a minute.

"Actually, just tie it around one of your belt loops."

"Huh? Why?!"

"I don't think it will fit around your thigh…"

"Am I too fat?" I said, sighing.

"Not at all. You… you just have a big bottom. You really fit tight in those nurse's pants. You must attract lots of guys."

"I do?!" I say. At this point I felt my face turn as red as wine. I panic as I try to cover my legs with my hands.

"It's okay, no one checks you out… anymore at least."

"What do you mean?" I question as I tie the darkened strip of cloth around my belt loop.

"One night, as you were sleeping inside the Humvee, some soldiers from the frontier army were stalking you…"

"WHAT?!"

"…and Will, when he was done with his night shift, he spotted them…"

"Did he kill them?"

"No," she answered quickly. Saying no more on the topic.

This is the first time I actually shuttered at the thought of Will. when Will, in my mind, was not Will but Iron Sights, the one-man army. A lost soul. Someone I have to help. And soon.

But I'll never refer to him as "Irons." Not in a million years.

"You're right to be scared, it's not…"

"I'm fine. No matter what, I still have a promise to keep."

September 25th, 2035

15:12

IH-45, Texas.

We parked on a hill that was four thousand feet from the ambush sight. Four thousand feet from the interstate of fallen comrades. And yet, from four thousand feet, I could see most of the soldiers with us still actively fighting.

But not Will.

"*Lower your head!*" Ivys silently hissed.

"I don't see him…" I sigh, quietly lying down and looking over at the half-dead trees. Well, at least trying to. This jagged rock did not like my chest at all, even with the field coat on. Luckily my legs were protected by these new Camo cargo pants that worked ten times better than the jacket. The OCP was tight-ish, but my maneuverability was not limited. I had to change anyway, and Ivys forced me to.

"Target spotted northwest. He's digging a foothold on his own. Amaryllis, how far do you think he is?"

When I poked my head out, I instantly spotted him under a dormant tree. Not dead or alive, not even thinking about it, I answered back as quickly as I could.

"565 to 570 meters."

"Copy that," she says. "Cover your ears."

I drop from peeking above the rock and cover my head completely as a shot is fired. The echo bounced through the dense dead trees. The ring of the shell of the bullet flying out sang shortly after

"Target down."

I begin to look up slowly until I notice the dead field. The target was shot, but the army continued to act as if no sniper shot was rung by them.

"Do they just not care?"

All I hear from her is "Cover!" before she grabs my head and shoves me down. I was so confused about why she was so rough this time.

"Sniper northwest!" she grits from her teeth, "She saw us. I saw her scope glare. We need to move."

"We can't! She knows we will be on the move, and she will shoot down more soldiers!" I say, poking my head.

"Target identified," I say, dropping my head half a second later. "1200 meters. Female. Abandoned radio tower."

"On it," she responds.

As she pokes her sniper out, I poke my head out to see that she is next to a water coolant system, conveniently on the right of her was a pressure chamber.

"She is too far to shoot any fatal shot, as so are we. Try to hit beside her, that tank beside her will leak loud, thick smoke, distracting her and giving away her position."

"On my mark, get to the car and announce on the radio to keep a lookout for the areas around the watchtower and to watch their heads."

I prepared myself to run, scared half to aspen itself.

But after a quiet minute, I began to run to the Humvee.

TING!

I get behind the door of the Humvee as a shot hits the body of the car. I didn't have time to look at the damage when I grabbed the radio and hid behind the dash.

"All units! Code 4D! Northwest Communications Tower! Watch your heads!" I yell before throwing the radio back and crawling out.

"*Cover your ears!*" Ivys yells, firing the shot.

It hit flawlessly. We couldn't hear the gas leak, but we could see the steam get pushed out like a white vape cloud as she scurries away.

"Can you hit her?" I ask, standing up.

"*Lay down!*" She hisses.

So instinctively, my body slams itself down while it is still super sore. Like how does she do this for *hours?* I painfully crawl over to her and again peek over the rock. I can see soldiers running through the dead trees, fighting and spreading out.

Looking for the sniper.

"I don't see Will still… I'm worried. Did he…"

"Relax. He's okay," Ivys reassures a ruffled and unsure response.

"It's getting late," Ivys says as she folds her sniper. The wear on it was intense, and the wood was close to giving her splinters. She got it at the last raid base, and she has held onto it since. I can tell that it was originally made for hunting.

Then she gets up and gestures for me to follow her. When we make it two steps away from the Humvee, I release two things.

One was why didn't we park farther away.

And the hole.

The blow that the sniper shot at us didn't dent the Humvee but punctured right through it. Straight through the steel armor and coming out the side fender.

Not even a 50 cal. It can do that damage. This was a cannon of a sniper.

"Why didn't we park away from our vantage point?"

"Didn't expect another sniper. Anyone downhill can't see the Humvee because of the angle, but that sniper had an eagle's eye view."

We got back into the Humvee and started it up. We drove down the hill and onto the battlefield, where thirty or so soldiers jogged towards us, all with maroon uniforms and shirts. I saw two heavily geared in black, but they both had a thick maroon strip tied around their thighs.

"Why don't we tie the strip around the arm?"

"Identify spies. Also, it looks too much like a Nazi armband," she answered as she stopped the Humvee and began to get out.

I follow behind and stand next to her as she stands in front of the Humvee, in front of everyone.

"Well, *someone* better catch me up."

"Ma'am," one of the ones in black started off. "We fought off the army and pulled them back north. We're still not entirely sure if they work for the sentinels or if they even *are* sentinels. They have gray breastplates with unidentifiable weapons. We captured eight units while Irons captured the sniper-"

"Will is alive!" I shout.

And they all looked at me as if I wasn't able to talk, and I just did.

"Y… yes. Anyways, they are held at one of their little camps, and we are awaiting orders."

"Drive. Everyone else get in, and whoever does not fit gets on the sides," Ivys commanded.

I begin to get nervous only because of the story I heard from Ivys. What if they were creeps, too?

"Don't worry Amaryllis. The rebels have sensible mindsets," she says to me. "You will be sitting in the front seat you're usually in."

Then she turns around and yells, "Please welcome Private Amaryllis to the rebel army!"
"HARROH!" They all chanted in unison, making me laugh a little.

And she was right. Before I even got inside, the man in black, who was told to drive, opened the door for me and closed it softly when I was inside. My seat was easily able to fit two but everyone who couldn't fit chose to hang on the sides. It was kinda scary awesome.

We drive through this dead-dense bush for about five minutes until we make it to this brown cameoed tent with holes inside the cover. Lined up were soldiers wearing gray on their knees in front of this tent, all being watched by a few soldiers around. So, we stopped, and another soldier opened my door.

"Thank you," I say as I step out onto the rough floor.

Walking up, I manage to spot Dominic, Monroe, and his "Goons," as Will calls them. I don't even know what that means.

Finally, I see Will.

"Will!" I practically scream, running to him as I slam into his stiff body and wrap my arms around him. Slightly tearing up.

"My daffodils… I missed you."

"Hello, Amaryllis. Not now," he says softly, pulling away.

"R…right." I pull away and look at his face that was filled with nothing but soot. No expression, no proof of life, just soot. His body geared and dirty while his ripped, gloved hands clutched onto his M27 for dear life.

"Let's make things easy for all of us," Dominic says, walking up to the lined soldiers. "Who do you work for?"

Every one of them stayed silent.

"Who do y'all work for again?"

"Fuck off, old man," says a heavily bearded man with a bald head.

Dominic unholsters his 44 magnums and cocks the hammer back, chambering a bullet.

"Did you know it's possible a human could survive with all his joints shot? Of course, they won't be doing much after that anymore," he says in an unserious tone.

"So… does anyone feel like singing now?"

Still, no one spoke up. All of them looked down, including the sniper I just noticed. In fact, she hasn't even lifted her head ever since we arrived.

"Every question that I don't get answered will result in one of each of your joints shot, so I'll ask one more time. I'm in a good mood today. *Who do you all work for?*"

"Trevor Boyten!" One in the middle finally said.

"Smart move," Dominic nods.

So, they are sentinels.

"And the *Beast* is getting supplies from someone?"

The Sniper girl flinched but stayed silent as her head stayed down.

So, Dominic shot the first bullet. The soldier, barely audible, grabbed his shoulder and landed on his other arm. Then he shot the next one in the same area. He yelled in agony as he gripped his forearm.

"WE DONT KNOW!" finally, the third one yelled.

"Pardon?" Dominic says, holding his gun up.

"Will, I'm scared," I say, and well… because I was.

"Shh, it's okay, they'll speak, and this will be over with," he says calmly. Collected.

"It's a supplier that only gives weapons and gear. He barely sent any troops with us. Beast won't tell us who they are. He gets mad and kills someone if we ask. The soldier that was sent with us only showed us how to use the gear provided. That's all we know, I swear!"

"Hm…" Dominic says, rubbing his chin. His harsh thinking was so quiet yet so loud. "Okay. You two-" he says, pointing at Monroe and… "Rose," as Will knows her by. "Take these two and treat them. When you do, put them in a truck and await my orders."

"Yes sir!" they both said together.

It took them a minute to pull the soldiers away, but now there were four left. And I was scared for them.

"Alright Men. Who is that one soldier?"

Everyone else lined up who was on their knees and looked up at the sniper, who still hadn't shown her face at all.

"Glad y'all are taking the easy route," Dominic says with a hint of satisfaction in his voice. "No more questions for you three. Y'all off the hook."

Soldiers from behind came up and escorted them up and away from us without command of Dominic. Leaving one and one only.

"Permission to take over?" I hear Ivys whisper to Dominic.

"Granted," he smiled.

Ivys then walked up in front of the sniper. She stops three feet away.

"Eyes forward, soldier," Ivys commanded. Looking down at her, but she didn't respond. The only thing that moved on her was her stomach for slow, steady breaths and her hair from the silent wind. Her hair was silvery, similar to mine; in fact, the only difference between our hair was that hers was straight and shorter. She wore a black unitard with gray plated body armor that practically covered her from head to toe. Dirty work boots similar to Wills, only gray.

Will walks behind a group of soldiers and comes back with a strange-looking sniper, I quickly inspect it as he hands it to Ivys. I couldn't even *identify* it. It looked similar to a railgun with two parallel rails shown top to bottom connected by a few curved pieces of carbon fiber. It looked like a blaster from a 2000s video game. The stock read in solid steel embedding "Lynx LR 50 Cal."

"This is not a sniper ever classified by any government. And this ain't some homemade science project," she says, inspecting it, reloading it.

"Who do you work for?!" She yells, shooting the gun into the air. It caused her to flinch again, but she still kept her head down

"Why can't we jus' kill 'er?" asked a blond soldier next to Zu. This must be the Beach guy "Troy" I read about. "She clearly ain't gonna talk."

"Stay away from this," Will says with an angered tone. "Fact check yourself."

"And you'd better watch yer tone, *Runt*."

Will then silently walked around me and right towards him. Before he could step up to him, I crushed my body to his to stop him.

"Relax, Will."

"So, this is the Irons energy I 'ear about, honestly 'm disappointed. Ripping out the eye of each guy that looked at this little bunny that night? Try hard much!"

"Will it be okay," I say. But in my mind, I could almost feel everything tremble within me. Did he rip out every one of their *eyes?*

"Admiral, control your troop," Dominic commands.

"Even if I commanded him to stop, he's gonna still continue to push. Trust me, I tried. I can't control if he hurts your kid."

"I'm warning you for *your* soldier's sake," Dominic warns. "Iro… Will. Back off."

After a solid one minute, he finally answered with a "Yes sir" and turned around. Walking away.

"And take your fucking bunny," he says.

Before shoving me close to tripping over myself. I couldn't see it but Will ran past me and tackled the soldier down before Ivys covered the rest.

"Don't look," she said.

All I hear is thumps. Loud thumps. Eleven solid hits. It sounded like a sack of potatoes smacking the ground over and over again. It took Zu, Dominic, and Ijin to pull Will away. The man was left in one piece, but he stayed on the ground, bloody and leaning up. Thankfully, he is still alive.

Bleeding was an understatement, though. His nose was almost caved into his face, his eyes were bloodshot, and his mouth almost drowned in blood. He spit out a bloody tooth, glaring at me and smiling with a red, sickened row of teeth.

Finally, when the sniper barely became visible to us, she started chuckling. That chuckle became a laugh. And that laugh became a full-on crazed bawl out.

September 25th, 2035

16:41

IH-45, Texas.

"*Look at you idiots!* I can't believe Silver's gear got captured by the likes of you," she says in a light German accent as she kicks her head back, laughing harder. "You guys can't even keep your army here together! Not even the legendary Iron Sights can keep his shit together!"

"This gear is by a *Mr. Silver*, you say?"

"And what's it to you, old man?" she takes back.

"It's everything to me, *soldier*," he says sternly. "What's the purpose of your ambush?"

"To fuckin' survive! *Duh!*"

"Judging by what I know about Beast's array of idiots, you guys don't need any more survival gear. Your skin is bright, and your cheeks are filled. You had a real reason to attack us." Dominic says before taking the sniper from Ivys and throwing it on the ground in front of her. "And it looks like your firepower is covered, too. In this case now we have nothing of value to you."

"You have your lives. Surrender now to the sentinels, and we will spare most of your lives."

"Right, and what the hell do you know about him?" Ivys asks, pointing at Will.

"Ohhhhh, the legendary Ironsights." She says, as if she couldn't tell who he was at first. "The *folktale* soldiers used to scare everyone. "*The perfect soldier.*"kills without command. Deranged yet unstoppable. A high-priority target, but we couldn't really identify him until today. That runt is feared by the Beast and most of his soldiers."

Will, staring at him but giving him no reaction, just continued tightening his blood-stained gloves.

That was impossible to believe. *How?*

"So, what *do* you want? We may just give it to you," Dominic asks, getting a death stare from Ivys.

"Three things, really. Each one for someone different," she answered, looking down again.

"Beast wants your head," she says, pointing slightly at Dominic.

"The army wants your head," she shifts her finger to Will.

"And I want the *seed,*" she says, dropping her hand.

Will walks away and goes inside his Humvee. He steps back out with his fist around the chain of the pendant of dirt and drops it in front of her, making her flinch slightly.

"*This?*"

She takes the pendent and inspects it. She and I could see the sprout had grown a little more than when I last saw it. Touches of green were hitting the solid white stem with a beautiful shading.

"The hell is this? M' talkin' about the boy."

"Jack?" I gasp.

Out loud.

"Is that the freak of nature's name?" she spits looking up. But strangely, her face lit with confusion when she looked at me as if I shapeshifted in front of her. Then her face hit with solid… joy.

"No one is leaving with you," Will answers, with Dominic's stern tone.

"In fact, you're not leaving either," Dominic backs up Will.

"*Laura..?*" The sniper whispers. She stares at me with her pretty black pupils.

"What did you call her?" Will asks, stepping up to her.

"Laura! It's me! Silvia! Can you not see me?!"

"I'm… sorry, but I think you are confusing me with the wrong person… my name is Amaryllis-"

"What have you *done to her?!* You *broke* her!" she yells to Will. "She was *Perfect!*" She turns to me again. "Laura! It's me! Silvia! What *happened* to you?!"

I stay silent as they pull her away. I manage to catch Will walking… almost speedwalking off where the… "*Silvia*" was.

"Wait! Will hold on! Dominic!" I yell, grabbing Dominic's big arm and turning him around.

"Dominic! They are after the boy I was assigned to, and they want to take him."

"Will!" Dominic shouts.

"Yes, sir?" He says as he picks up the golden pendant from the floor, dusting the clear vile off.

"Your new assignment is to protect the boy with Amaryllis."

"What?!" We both exclaim.

"Understand that they want that boy as much as they want me or you, maybe even more. You are the soldier here most capable of handling that much protection detail, all I'm asking is for you to watch your brother. Is this too hard of a job?"

"No, sir."

"Good. Amaryllis. Inform the boy of what's happening."

"Y- yes sir.." I stammer.

Finally, Dominic dismissed us. Will afterward silently walked back to the Humvee and got in the driver's side. I follow behind on the passenger. When I close the door, finally, after weeks, Will does the talking first.

"You know you're gonna have to tell Jack-"

"No, *you* tell him. How would I be able to tell him?" I ask with a *very* confused tone, just to get my point across.

He puts his gloved hands together, interlocking his fingers while closing his eyes. As if he was praying — very stressfully praying.

"Have one more session with him. Alone."

"For what?"

"I just want you to find out why he hates me specifically. I need you to ask him before I approach him."

"I don't know Will-"

"Amaryllis, *please*," he begs in a soft tone I have never heard before. Surprising me. Convincing me.

"Okay, Will. I promise to try."

20:16

Today, a dozen frontiersmen and two of our men died in battle. What is a small dent in our numbers was still catastrophic for us only because a soldier is still a soldier. A life. Their names were Blaze Socrates and Jay Pines. We cremated them when we got back.

Now, I stood in front of Jack's truck with worry. Mostly from what task I had next but there was a hint of worry from the flowers sticking out the cover of the tent door on the truck. It's becoming increasingly difficult to hide his ability from the common people. Even Will knows about it at this point.

20:04

"He has an uncanny ability to grow flora from his persons and body. Studying further I discovered that Jack has a specialized strain of photosynthetic bacteria that reside inside his skin cells through a complex biochemical process. Regular nutrients he takes generate energy for promoting growth of roots and flowers as a natural extension of his body."

"How did you figure *that* out?"

"Aha. um…"

So, I don't know if I really do want to tell him that I was digging through his bag reading the paper binder written by… *Laura Silvers.*

She didn't sound familiar. Not by name or actions. But Silvia, on the other hand, was mirrored in style. She was taller but slimmer. Her hair was in similar silver designs as mine, but it was straight, and her eyes. She had keen hunter eyes, similar to a hunting animal.

Was I a bad person? Is what Silvia said true?

Am I *Laura Silvers?*

Someone who conducted horrendous experiments on a child for years without his consent? Calling him nothing outside of "Experiment," "test," or "Seed?" never by his name… even the experiments vary from heavy physical and strong mental challenges. And when Jack was quote on quote "Upset" with these "Experiments" he was labeled "rebellious" and "Mentally incompetent."

I would never do that. I *could* never do that. Maybe Silvia really did confuse me with a coworker or possibly the real Laura.

Rather so. I can't tell Will this yet.

"I… got into some research."

Technically, the truth, but again, technically not.

"He could be heading to bed now. Catch him before he does."

"Of course."

"And… hide those vines…" he says, staring at the flaps of the truck door with flowers peeking from the corners.

We were parked a few feet away from the actual front of Jack's truck when I got out and walked towards the front flaps of the inside. Finally, I was able to build the chest to knock on the solid steel doorway.

The opening unzips, and Jack is there, wearing nothing but pants, rubbing his eyes. Seeing him like that I understood and could see why doctors confused his condition with a disease.

At around six random parts of his waistline to his neck, arbitrary roots sprouted out. The surrounding skin, about an inch from the actual root, was turning a darkish green and purple color. He must have just trimmed the vines off because he was able to somehow throw off his shirt.

"Amaryllis… hello. Are you okay?"

"Yes, I'm okay. I hope I'm not disturbing your sleep."

"Nah, not at all," he says, flicking his wrist as he yawns, making me let out a small laugh.

"Come inside," he says, holding the tent up for me even though he isn't really the height to be doing that. When we get inside, he throws on his shirt again and sits in a corner of thick flower beds. "Is it you or me in trouble?"

"No, no, no one is in trouble." I begin. "I just need to bother you for one thing. But I will warn you, you won't like the talk," I sigh.

And instead of sighing with me or even complaining, he sits cross-legged in his bed of flowers and pats the floor in front of him.

So, I sit on the vines of flowers as they slowly move past my legs.

"I'm trying my best to control them…" he says as he sees me look at the vines below.

"You're doing a good job. I would usually be wrapped up like a spider now." I giggle. "Practice makes perfect. You're getting better by the day."

"Mhm," he says, reading my face, anticipating me with his deep hazel eyes similar to Will's.

"Your brother. Why don't you like him?"

His small grin began to fade a little as the petals of the flowers around him began to close.

"Jack, I-"

"Who said I *didn't like him?*" he says softly. Almost heartbroken. "I love him, As ya know… brothers. I love him so much, and I don't avoid him because I *don't like him*. I *do* like him. That's why I avoid him the most out of everyone."

"*Why?*"

"Look at me. I grow *weeds* from my *body!* I have vines that grow so long and that are *hardly* controllable. He doesn't even need to see what I am, the actual… monster I have become." He pauses for a moment as the petals of the flowers close, and a tear rolls down his cheek.

"I even asked Mom if she could keep him away. He already had a lot to deal with at home-"

"Jack! You know he *begged* your mom almost every week to see you. Your brother has traveled miles to find you. He knows about your special ability. In fact, when he found out, he was barely fazed."

"What?!" he shot up. "He *Knows?!*"

"Yeah, he does. And even with all he knows, he still cares so much for you even though he thinks you hate him with all your guts."

"*He does?!* No, not at all! Please bring him here whenever you can!" he says. His grin slowly grows as the pedals of the flora around open back up.

"He's outside. He needs to see you, too, anyway."

"Can I go get him?!"

"I'll go get him. I don't want to get in trouble for you sneaking around at night," I say, getting up and walking out. I could hear Jack scrambling around in his room as he tried to make things… presentable.

"Will!" I call out to him as he leans on the frame of his Humvee.

"Is he ready?" he asks, walking up to me.

"Yes. Are you?"

"No, let's go," he says, walking to the truck bed.

When we made it to the soft door, I stopped Will and shouted to Jack, "Ready?"

"Yeah!" says a muffled Jack inside.

Will flips the door flap up for me to walk in while he follows behind. And wow, I didn't see him well in the dark.

Of course, he towered over us as he laid eyes on Jack for the first... well- the second time in five years. His Uniform Camo and dirty gear that covered him, combined with his overgrown, very dirty hair and dirty face, made him look almost unrecognizable from when I first met him.

"Hey, Jack... sorry for the mess- darker colors blend better in the environment..." Will says in the soft voice I only heard once before.

"Bro, Will..." he starts as the flowers around us start to open like a bright summer day.

"You look just like dad."

September 25th, 2035

20:44

Huntsville, Texas.

I left the tent shortly after that. I figured they deserve the alone time after all these years. I don't know if anything special happened after. If anything did happen, I wasn't there to see it; I was outside, barely keeping myself awake.

Finally, Will came out to get me, flowers weaved around his legs clung to him slightly as he peeked his head out of the door.

"These won't kill me, Right?"

"Relax, Will, they are harmless." I giggle, walking inside. Jack was sitting on the ground, folding an envelope, which I managed to read the bottom words of.

WITH MY FINAL GOODBYE. YOUR DAD, THOMAS LIGHT

Fort Tundra.

Dec 12. 2034

"Are you okay, Jack?" I say as I see him teary-eyed, looking down at the folded envelope. Yet the unique thing about these tears was that they didn't close up the petals of the flowers around him.

"I'm just so happy. Mom is okay… that Will is okay, and I'm finally feeling free. These days can't get better; even if the world magically fixed itself, I will still say that tonight was the best I ever had.

And will respond by just sitting next to him. *Will.* he had the face of Will a smile that I had never seen before. A clean, soft tone, That's Will.

Not *"Irons."* Or *"Ironsights."* Will.

"Are you guys caught up?" I would ask while they talked.

"Of course not!" They both would answer, laughing.

While we made Will a flower crown, he broke into conversation, starting with, "By the way, Amaryllis, if the place you say is real, you understand that we will be moving a lot of people, right? There was our camp which had a bunch of civilians. We will most likely bring them there.

"And live happily together! With mom!" Jack cuts in.

"Yeah, with Mom," Will says, giving Jack a Noogie as Jack tries to pull away playfully.

"It's getting late, boys. I do feel as if we should be sleeping," I say, smiling slightly.

Only slightly, because the night will soon end. The moon… I sometimes miss the pretty flower field in the sky that it brings with it. Now, it's all thick and dark. No stars, moon, or even sky.

"I guess I should do some night watch then," Will says, getting up with tightened vines around his legs.

"What I mean by "*we*" is that you should go to bed too," I say, trying to sound as firm as Ivys.

I didn't do a good job because all he did was ignore me as he pulled himself out of the vines.

"Yeah! Please don't go! Sleep here tonight?" Jack begged, standing up with him.

"Jack…" Will starts, but seeing his face makes him read enough to know that there is no point in arguing.

So, he sighs. "Can I just *walk* her to the car? I need to drop my bag there. And I need to make sure she eats."

My stomach growled at the sound of that, but I maintained my composure. I didn't even know I was hungry.

Jack nods. And… loosens the vines around Will.

"Your control is getting really good. Keep practicing and I promise you will be allowed out with everyone."

"I don't know why I can easily control it now rather than before. I practiced every day in the hospital, but I could barely even move them."

"You were built around *fear* before. The emotion stunted your control over your growth, just like in any other plant. But now you're calmer… the serotonin in your mood is making things easier to feel. To control."

"So… because I'm happy? Jack asks, sitting back down.

"Yes, because you're happy."

"Well, go!" he says, shooing me and Will. "Go! Go to bed!"

"Alright," I say both in my mind and out loud.

I hear him yell out goodnight when we step out of the truck. Before hearing him shuffle around and zip up his door again.

"You know, I may be wrong, but I think he was happy to see you," I tease.

September 25th, 2035

10:31 PM

Huntsvill, Texas.

To be honest, thought I burned this book already. Started it a day *before* the world ended. (*Coincidence?*) Plus, I wasn't really in the writing mood at the time when the days were going by.

So, it really surprised me when my night ended with Amaryllis giving it to me after we ate a nice warm MRE. It did seem hard for her to let go of it. But she did while giving a twisted question.

"Will, am I a bad human?"

"*What,*" I answered. Shocked. Like. What.

"Why would you *think* that?"

"I just- I don't know who I even was before. Before I forgot everything about myself… the sniper from the field said that my name was Laura Silvers. An Epomologistic doctor who studied Jack in Horrible ways. Hosted torturous experiments on Jack and his ability. The one who took you away from your brother, gaslit your mother, and scarred Jack as nothing but a test for herself. What if I am them? What if I did that to others my whole life-" she says, tearing up.

"Woah woah, what are you talking about?" I ask, grabbing both of her shoulders and turning her around, holding her chin up to me. "So, what if you were or not? Look at who you are now! Who are you?!" I say, lightening my tone as she looks up at me with her big eyes.

"I- I'm Amaryllis," she says softly.

"Exactly! Whoever you were before, even if it was Doctor Laura that took Jack, you're *Amaryllis* now. And only *Amaryllis*." I say, hugging her.

She was small and fragile. It was hard hugging her softly with my weight. But she takes it for about a minute before pushing me away softly. Smiling.

"Thank you, Will," she says, getting on the tips of her toes and kissing me softly on my cheek.

"Now go," she says as I stare blankly at her.

"Um… Will. Jack?" she reminds me.

"Y- yes. Of course," I say, turning around, tripping on myself occasionally as I walk away.

Ohmyfuckinggodshekissedme?! Is all my mind kept relaying over and over like a bad song stuck in my head? And I try my soul's hardest to clear it when I see Jack.

"Bro, why are you smiling so much?" Jack asks as I stumble inside.

Clearly, I didn't manage my smile well enough.

"Something lodged in my mind," I answered, turning off the dim tube light. Dropping the subject.

"Thank you for staying tonight," he says, lying down seemingly comfortably. "I'm sorry for begging."

"Yeah," I answered. The reason I say seemingly is because when he lays down, he feels as comfortable as ever. But I try laying down on the solid barley padded ground, and it's worse than the grass *outside*.

While I was twisting around, trying to get comfortable, the flower vines thickened below me and lightly wrapped around me. Warming me and bedding the floor.

"Thanks…" I say, surprised.

"You're welcome, Sleep well, Will," he says from my side.

"Night Jack."

September 26th, 2035

7:21 AM

Huntsville, Texas

The morning came quicker than expected when Amaryllis was the one shaking me awake. The Mayor Alder and his assistant… (which I never learned the name for) woke up Jack. It was hard to climb out of the vines, but Jack managed to help a lot. It was even harder leaving him again. But I had to get up fast. Amaryllis notified me that the commanders had news for the Army and that we needed to round up.

So, the morning went quickly. I ate breakfast with Amaryllis and grabbed my stuff. And went into a formation that was built in the middle of a clear courtyard.

And so, when everyone arrived, the announcements were summed up into 3 parts.

"We are about two hundred strong. Our units consist of six Tanks, four mounted Humvees, ten stock Humvees, And about a dozen motorcycles," Colonel Dominic announced. "Not including the four buses we have for the uncounted civilians."

"Buses" means the two prison buses and the other two carrier trucks,

"So, with these numbers, we will begin to make our way to the bunker. Once we arrive, we will split troops to bring any and all civilians to our last base of operations," Colonel says, bringing light to the third topic.

"And, of course, you all are probably wondering about the enemy forces. With our troops fighting an unknown supplier for the Sentinel army… Well, the plan and mission continue for us. Defend and push through 'till we are safe. Until we discover who is supplying the sentinel army, we keep on our line. And keep ourselves from getting killed."

"Yes, sir!" everyone chanted.

"Now then, I need a group of twenty soldiers pumping out gas with me at a gas station down the street. So, these two rows, follow me. We leave at dawn tomorrow. *Fall out!*"

8:33 AM

I was busy loading up tents for the civilians when the First Sgt appeared. He began questioning me on why I didn't take the night shift. However, his questioning was not in a challenging way but more in a curious way.

"My brother wanted me to spend the night with him."

"Oh-" she says surprised. "id forgotten that Dominic tasked you with that. There was no fighting?"

"Why would there be?" I asked, as surprised as her.

"From what I learned, time really showed that he didn't want to see you at all."

"Misunderstanding, ma'am. He and Amaryllis were just trying to protect me and everyone else."

"From the disease?"

Right. Only a certain few people actually know about Jack's ability. I mean, we aren't keeping it as a secret per se, more like a *"make sure as few people possible know."* Kinda thing.

So yeah, I treat it as a secret.

"That's one of the reasons, yes."
Again, technically, it's not a lie.

"I have noticed a change in you when I see you after your trips with him. A good change. I mean, a day ago, I thought I was never going to see you smile," she says, leaning herself on the table. She takes a loose strand of her hair and sticks it behind her ear.

"Can you do me a favor?" she says in a lowered tone, surprising me more.

"Huh?" is all that escapes from my throat as I drop one of the tents a little too fast.

"Can you *please* keep Amaryllis safe?"

Honestly, I thought she was about to tell me something harsh, like *"Can you please wash yourself?!"* or *"Can you not slam the tents down?!"* and make me do millions of pushups.

"Y-yeah? I don't understand why you're reminding me. I already try to. Am I not doing a good job?" I ask, hoping I didn't break the promise I made long ago.

"You're doing fine, and I just hope you know that she saved your ass."

Yeah. I knew that silently.

"Of course."

When I finished up, she was already gone, so I went back to the Humvee and dropped my bag off again. Noticing the gleam of the pendant hanging from my rearview mirror glowing on me, the seed inside still full of life.

It started growing then. Whenever I snuck away from any of my posts to be with Jack. In a way, all I did was protect his surroundings. But just that distance away from him, along with the time I was even with him, had been enough to cause the seed to sprout. Whatever Jack has is something else entirely.

Even Amaryllis. Her ability to know anything about it speaks from the perspective of the last person she was before. A hidden secret from her past that no one knows about. Not even herself.

Unless there was one who *did* know her.

September 26th, 2035

11:15

Huntsville, Texas

"Well, if it isn't the legendary Ironsights himself! What do I owe the *honor?"*

Silvia, her pain-in-the-ass sneer, tells me that she will be trying to con me for info as much as I am her. She wasn't the only one we have captured. Another Kraz truck similar to Jack carried the rest. Only the automatic door was replaced with a welded barred door. The driver and warden of these soldiers is named Lugar. Appointed by Zu, he knows how to do his job. Despite the Frontier and even some of the Rebel army wanting them buried, we have managed to keep them as fed and taken care of as any civilian with us.

Partly.

And I say that loosely because they do only get their food and water via cans and bottles that are loosely thrown inside.

They are prisoners of war. We are treating them like *royalty* compared to what they would have done to us if they captured us.

I managed to convince the First Sgt to let me speak to her, after a little effort, for one reason mainly.

"We already tried getting intel out of her. All she does is banter about random bulls. She won't budge. Go ahead and try," First Sgt warned

"I have my way," I answered back.

My way was just to wing it.

So, we pulled her out and brought her inside an abandoned store. The two exits that the store had were guarded each by a frontier soldier. And an old heavy freezer guarding the third.

And so there she was, sitting in front of us, cuffed to a plastic lawn chair. Her hair shrouded her face as if we were back on the field.

"Who is Laura?"

"My ass, kid," she snaps back. Kicking her head up so I could see her. "As if you didn't know already, she's acting."

"Acting," I repeat.

"Swear' to whatever god you believe in. She is still a bitch of ours."

"Ours?" I question

"*Kid*. I don't needa tell you shit, not you or any of the folks around. As soon as I do, you will kill me off. Ts' that simple." "No one is gonna kill you," I assure her.

"Now that's bullshit."

"So… "Laura," as you say, she actually did lose her memory. How would you even know if she is her?" I ask.

"Don't act like you've never seen the "one" on the back of her neck."

When she said that, I flinched. I remember not even worrying about it because it was just a tattoo and nothing more, I had so much already on my plate that it never even crossed my mind.

"You know *exactly* what I'm talking about," she says, a tone cold and harsh.

"What can I learn about Laura? Because she is gone."

"That bitch died?!" she shouts.

"No! She is now Amaryllis. She would have broken you out about now. I can bet that the only thing she is doing right now is chasing flowers." I snap back.

"What's in it for me?" she sneers again.
"What do you want?"

"Nuthin' you can give."

"You want to leave?"

"Great thinking Detective," she says.

"What about a different compromise? Do you want to see Amaryllis?"

"That spoiled *bitch* can fuck off," she says, spitting on the ground in front of me.

"*Spoiled bitch* hm?" I repeat.

"That *Daddy's girl* treated me like I was a worthless *mongrel* to her. Commands left and right. So smart yet so *annoying*, just like our father!" Silvia yells. Kicking a rusty can that was beneath her chair.

"She is your *sister?* Why do you hate her so much?"

"Hate all my folks. My father is ten times worse than her." "Amaryllis isn't like that. She is quiet, honest, and helpful to everyone! She knows Laura; she understands that she could have been Laura, but that life is gone now," I almost yelled. I would be surprised if the guards didn't hear me.

"Even if. Our father will still kill her and me as soon as we return."

"Then why go back?" I ask, astounded. A father ordering his kids to death points like they are servants. And the kids don't even *mind* it.

"Every *second* I don't return is a slower death for me. He will eventually catch me, as he has before. He can and he will."

"If you go back, you're dead no matter what. Stay on the run. With us. Then you will have a chance of living," I said softly.

"A very low chance."

"A chance nonetheless. With you alive, you may see the sun rise again," I say, getting up and loading my gun on my shoulder. Clearly I'm not getting much from her, so I'll slow down trying. She already dropped her head down again.

"Beast's supply is being funded by a supplier known as Omen Silvers. You ain't getting any more out of me."

I walk up to her and pat her shoulder, hiding my huge surprise. "Thank you. We will keep you safe."

"I didn't ask. Just promise me something."

I look at her, anticipating what she could possibly hold against me.

"If things fuck up, kill me. 'I would rather a quick death than one from my father."

I walk back to the frontiersmen and nod to them, signaling a finished conversation.

"You're not gonna die. You have my word."

"don't worry. You will" Silvia answers back.

September 26th, 2035

8:01 PM

Huntsville, Texas

I learned something tonight. It's that I hate telling Jack no.

"*Pleaseee!* It's our last night here!" he begs.

"Sorry, bro." It pained me to say. "Duty calls. Gotta keep the night watch. If you want, I can bring Amaryllis."

"Amaryllis went to sleep already. I thought you would know because your car is her favorite bed," Jack says, pouting, as he crosses his arms.

In case you couldn't tell, we were back inside his truck. And he wants me to sleep over. One last time.

"*Actually,* I didn't go by the Humvee all day. I got busy," I answer.

"You know, it sucks you can't sleep in, it's only been like *five years?!*"

"C'mon, bro, go to sleep. When we get to the bunker with Mom, I promise I'll never leave again," I say while grabbing my M27 from the truck floor.

Then I kneel down and pat his shoulder. "We are brothers. From soil to flower, I will love you forever."

"Fine," he caves.

As I walked out, I could hear him whisper, "Don't get shot."

The night went on slowly. The atmosphere was cold and dark. It clung to my skin as I kept my eyes out onto the land around me. Not even my gear weighed me down, but my thoughts did. Complex equations filled my mind with stories.

Amaryllis.

Silvia.

Trevor.

Omen.

All these stories mixed together in a picture that was in a message un-transcribed. All of them are linked like a modern art masterpiece. Chaos and little order are shown somewhere in every single one of them. The Beast has a funding officer who could be Amaryllis' father. Silvia's side of this story shows me that this father is very keen on the idea of killing his own family.

What really weirded me out was the numbers.

Counting *them*. If that's the case, how many are there?

The tattoo clung to Amaryllise's neck, and I can bet that she doesn't even know about it.

Or does she?

Omen. Whoever he is can and will hurt us.

Colonel Dominic needs to know. He needs to know a lot. Beast gets closer every day.

I moved away from my post two hours after I got there. Going to where most of the vehicles are parked I could see my car past the line of whatever else had to be fueled up. Peeking my head on the hardened glass, I can see her silently knocked out.

Yet Dominic wasn't there, so I continued moving down the line of lights and engines. And I do see him packing up all of the pumping supplies with his boys. Six of them, which included him, were all illuminated from the headlights shining on them. Of course, he was in the middle, taking charge until he saw me. He walked away from the loading and up to me.

"Iro.. Private. Can I help you?" Colonel says, as stern as he usually is.

"Intel, sir. The supplier for the sentinels is named Omen Silvers. That's all I can say. No location, no destination, nothing extra, sir."

"At ease, a name is enough. If anyone here knows him by memory, then we have intel. Thank you, Private. Fall out," he calmly says, saluting.

I salute back and walk back to my Humvee. I open the driver's side door and almost shit myself. I didn't notice it at first, but in the passenger seat in front of Amaryllis was Ivys, sleeping solid and quiet.

At only 10:00 pm? I thought Ivys *hated* Amaryllis. Or at least didn't trust her. But from what Ivys told me recently, and her sleeping with her shows that something changed.

I guess she is warming up to her.

It's smart to sleep in anyway. Tomorrow will be our last non-stop trip to this magical bunker of freedom.

We are also splitting off the escort tomorrow to send off to the School base to pick up everyone else.

Colonel putting this much trust in Amaryllis without even getting to know her shows that he must be desperate. I hope Amaryllis is right. Trust is easy to lose.

I'm sure Dominic is gonna strangle her if she is wrong, which scares me even more.

I never found anyone I could truly trust before. Rather my mother but even *she* lied to me. However, there was a time when the closest person I trusted was my father. Especially the day he taught me how to use a gun.

June 7th, 2031.

3:44 PM

Abilene, Texas.

Of course, it started with him coming home after basic. A few months after Jack left. I was still thirteen, and it was close to my birthday. My father wanted to give me an early birthday present before he got deployed.

So, there I was, holding the heavy 1911. The concept of a gun itself was foreign to me. From the sights down to the grip, I couldn't even fathom how the tool I was holding worked.

"Dude, what are you *aiming* at?" he said in a mocking tone, the one I used way back.

"At the bottle, obviously?" I say, gripping the awkward handle tightly.

"First off, that's not how you hold it. Put your other hand on the bottom of the magazine."

"What? Like a newspaper?"

"No, no…" he chuckled, "the bottom of the handle. Like you're holding a candle."

I followed him word for word. And the handle seemed so much easier to hold. Aim and grip, and I was surprised at how comfortable it felt.

Now, the problem was keeping the gun still.

"Good. Now, aim with one eye. Close the other if you need to-"

"Wait, when you shot the gun you didn't even close one eye! How come I have to?"

He again laughed at my impudence.

"You really are a visual learner. It's a hard skill to center both eyes with what the gun is aiming at. But once you learn it, aiming almost becomes second nature."

"So, teach me that!" I complain, keeping my other eye open as I aim.

"Sorry. You're going to have to start easy. Most guns are scoped now, so I am just showing you how to use them.

I pouted and sighed, but he was right.

"Alright… it's easy to aim, but how do I hold the gun still?" I ask as I try to keep the little line of metal onto the bottle of water.

"That should be the easiest part. *Calm down.*"

"But I am calm," I say, still trying to aim down. Trying to control the violently trembling gun.

At this point, I couldn't even tell if that was me or the gun.

"You're imagining it as a gun. Don't. It's just a tool. One that you respect and take care of. If you do, it will take care of you."

I nod, pausing and listening to every word of his. The wild mare of a weapon was starting to steady.

"Good. Now breathe slowly, in and out. Slow your air with every breath."

As I began to follow, around the third breath, my mind started to clear. I couldn't hear my thoughts of my mother Alice if she would see me now, or my dad's reaction with her. Whatever happened in the past and whatever happens in the future cleared my mind, my focus was aiming. Perfect.

BANG!

The sound was scary, surprising too. Expected a firecracker at best, not the screeching pop of brass exploding in my hand that I heard. The recoil shot the gun up, and if I didn't grip it way too tightly, it would have ended up behind me. And it was for nothing. Looking over, it was a damn miss.

"I can't do this. It's impossible, I'm too small." I complain, wiping my sweaty face. I grab the barrel of the gun and hold it out to him, waiting for him to take it.

Of course, as well as you and I know my dad. He didn't take it. Instead, he taught me another skill. He corrects my gun and puts it back to aiming at the bottle.

"Every skill grows for how much you feed them. If you aim for nothing, you will hit nothing. Let's try again. Never give in."

So, I sigh, aiming again. Centering my eye with the iron sights onto the bottle again.

"Breathe slowly."

Again, after only a second breath, this time, my mind clears. The only thing I focused on was the bottle, my shot ready. My mind cleared off of only one thing.

Never give in.

Ready.

Aim.

I pull the trigger once my surroundings go black, and I see nothing but the condensed water bottle.

BANG!

CRASH!

The tinkling glass flew all over the area around it. The satisfying landing of the bullet hitting the ground behind it with a thump followed slowly after.

"*Yes!* Did you see that?!"

"I'm right here, dude." He chuckled. "But we are only ten feet away. Let's move another ten."

"Alright!"

September 26th, 2035

11:00 PM

Huntsville, Texas

We shot a couple more rounds before ending that day. If only I knew that was a good memory. I was so lost in the moment, taking so much for granted. My father, The plants around, even the *sun*. But that day, I learned one quote.

The one quote that stood out to me the most of any.

"Never give in."

No matter how absurd the circumstances are or how repetitive the quote becomes. We have to remain focused.

September 27th, 2035

7:10 AM

IH-45, Texas

The night ended well. No ambushes or spies or anything down that line. When everyone woke up, we packed up quickly and easily. The Colonel has taken the route to send back two Humvees with the buses.

Amaryllis better be right.

Ivys took the wheel today. She really insisted so that I couldn't go against her. Only five minutes down the road, Amaryllis starts leaning on me again, trying to fall asleep.

Of course, I didn't stop her.

"Get the plan, Will?"

"Yes, ma'am," I said firmly after she explained it. To (again) put it in "Will" terms, about a third of us will be split off. I will be going with Ivys, the doctors, a few soldiers, and Amaryllis to all find this silo, or bunker, or whatever. The rest of the troop will head into deeper parts of Houston, down Bay Town, to this Naval base to get our final clearance to send the emergency SMS announcement country-wide. We will regroup with them after the scouting. If we are lucky enough, we will have a radio connection. But now is the waiting game.

A six-hour waiting game. And the trip didn't change much. Monroe still sat on my right side… *smaller?*

Yeah, no, he's still taller, but now there's… a difference. Have I gotten this much bigger? Gotta check when we get out next time. Besides, I have someone resting on me right now.

"Sir- Look west," the First Sgt says with a new tone. And, of course, all of us looked.

Grass. The flooring of the far open field was littered with slow-growing patches of green, sprouts of a new world growing past the decayed one. It almost looked like seeing stars for the first time. The grass growing with the lack of sunlight and moisture really shows how tough nature is. Colonel grabs the radio and starts speaking as we drive past the field.

"Ignore it, lads. Nature is taking its course that we shouldn't interfere with," he swiftly spoke, dropping the radio back in its place.

"It will be slow, but the world will be reborn on its own." Colonel sighs.

Clouds covered the path forward in a dark tint. Visible and dark, and yet we still kept driving. Driving for the next hour was quiet, aside from the engine roar and the rocks of the carriages of steel and armor. Amaryllis slept through it all on me. I need to learn her ways.

"Split off section. Scout team, Take the turnpike." Colonel says on the radio. Before the Humvee halted with everyone else, Colonel stepped out and looked back inside. *Take command, First Sgt."*

"Yes sir," she responds, moving to the driver's side. Colonel nods and closes the door after a quick goodbye.

We are on our own now.

As we get ready to move, she switches the Radio to channel .52.

We were given four Humvees, two of which are mounted. We also got two motorcycles to accompany us. A small army but an army of Rebels. Which means we will have the best coordination in case of anything

Oh, and Jack's truck too. So, he's coming too. That's the only reason I'm scared to split up.

"Will! Wake her up. I want her awake so if she is wrong, I can kick her ass."

God. Amaryllis, you better be right. "Yes, ma'am"

And we start driving again, leading our own line of soldiers.

"Amaryllis..?" I said softly.

Note this. I only said that. I thought she was awake already and was just resting from how quickly she woke up, but it only took a word of name to wake her up. I didn't need to tap her, shake her. I didn't even need to *touch* her.

"Mm... ~Hello, Will." She yawns, looking up with her pretty brown eyes.

"We are almost at your Bunker you told us about-"

"-And you better be right." First Sgt cuts in.

2:10 PM

We are about ten minutes more down the road, reaching the fallout shelter as stated by Amaryllis. There are factors we do need to consider, even if it is there, of course. The term that it is there, how will we be able to access it? What if it's being used? Or even the small chance that it was destroyed.

"That's all on the account that it's actually there," Ivys says.

"I promise it's here. I feel a familiar feeling about the area. On our next exit, you will see an empty field. We are pulling onto the downhill viewpoint."

We drove, tense of what we would see. Me and Monroe were already bringing our guns to our chests.

Did we trust her? I did.

But we didn't.

But a few minutes later we take the next exit onto a bridging street that leads to a hill. And from where we parked and halted the brigade, on that hill we could see the clearing of the field.

And there, a foundation stood on the empty lot. A dome of concrete stood up not too high but not too low to be classified as puny. In fact, it was large. As large as a barn. A circular dome stuck out like a pin with a slice in it as if it was cut like a pie. The sliced entrance lowered into the ground a little and onto a huge wall, aka the door.

It was real.

"I'll be damned. You were right, Amaryllis…" Ivys says, taken back.

"Let's go then!" I excitingly rub my hands together.

"Not all of us. It could still be dangerous. Me and Amaryllis will stay behind, and so will Jack. The doctors will go with you, and they will leave Jack up here with me and Amaryllis. We have to keep things safe," First Sgt said.

With this plan, she will have a birds-eye view. Plus, she will have a sniper. But will it still be enough to be classified as safe..? At least to me? I weighed the options, coming to the terms that this *was* the only option.

"Okay."

She grabs the radio and explains what's happening after we switch. When she gets out I hoist my bag on my back and climb to the front as Amaryllis follows her. Monroe follows me onto the passenger side. Soon, Ivys and Amaryllis step away, and Ivys comes back with Adonis and the nurse hopping into the backseats.

"Will," First Sgt taps my window.

"Yes ma'am?" I answer after I lowered my window

"Be safe. Usually, I would wish you luck that you don't get attacked, but I'm dying to test out this sniper," she says, lifting the railgun-looking device. The one that Silvia had. Lynx.

"Understood, ma'am."

"*Oh.* and take charge."

Take charge…

Take charge?!

"E… Excuse me, ma'am..?"

"Are you contradicting an order?" Ivys shoots back.

"No ma'am! But-"

"I nominate you as First lieutenant. Your officers are Ijin and Monroe."

"Y… yes, ma'am."

I wanted to argue, but I literally had no other choice.

I'm in command now.

"I'll be connected to the same radio station as you. If anything, you can contact me."

"Yes, First Sgt." I answered, but secretly, my head was like, *"fuckfuckfuckfuckfuck"* I'm a *commander* right now. For a real military army.

Before I hesitate, I roll back up the window and start driving, each of the army following me.

"Irons. You got pressure on you," Monroe chuckles.

"Getting past that."

"With what I've seen from you, you do have the capacity to lead."

"Yeah, but… an *army* is a little much, ya know. I'm fucking seventeen," I shot back

"Who the hell asked how old you were? I'm 21, and do you see me aching from life?" he asked harshly.

"No?"

"You're now in command, kid. All these soldiers rely on you, rather you take action, or you shut up and continue *whining like a bitch.*"

Yeah. Getting past the "tough talk," he was right. I thought I was gonna hate this asshole for life.

"Let's just get this over with."

"Atta' boy," he says, grabbing my shoulder.

"Turn onto the field here," Adonis says from the backseat, pointing out a smashed fence from another meteor leading into the field. We squeeze past and finally start driving out of formation until we reach the solid concrete structure.

"Hold position," I say as everyone halts at once. In fact, I think I was the last one to stop.

My mind racks with heat as I form a plan quickly.

"Okay… um… Man, the guns and stay by your vehicle. Monroe and Ijin, follow me to the door." I say, dropping the radio back down.

"You guys, too," I say, addressing the doctors in the back.

"Coming along," Adonis says, stepping out with the nurse whose name I can't remember for the life of me. Either way, I try not to look directly at her face as we get out.

Funny enough, the troops listened. Ijin was walking up with an M16 in hand. A full dark silver covered it head to stock.

"Looks like the newbie became a commander. It's less surprising that it's Ironsights."

"The door," I commanded, ending the talk from that point.

Silently, we walked up to the wall of iron and concrete that was supposedly supposed to move. Twelve feet by twelve feet, this wall stood in front of us. In black paint at the top was a large "EU-12."

"Stands for *Evacuation unit.* There might be people inside," Adonis says, walking up to a black screen beside the door. An iPad-looking device embedded itself into the wall.

"*Hand scanner,*" Adonis says, spreading his palm on the black screen. A second later, the black screen flashed a blue light on his hand at rapid speeds before it flashed red silently and showed a red X. Below it, read:

ACCESS DENIED

"Guess I'm not one of the big shots," Adonis sighed.

"So, who is?" Ijin asks.

"This bunker was made by ECTO, so it's funded by the government. Probably some high-ranking official, like the president or something."

That's lame. We can't brute force into this. A damn meteor shower couldn't. We are locked out unless we meet someone or find the body of someone who is high class in the government. Who probably are all rather dead or locked up underground already.

"*Pfft,* the president is probably locked inside now, getting spoon-fed with a golden spoon," Monroe yells, laughing loudly.

Yes, the United States' first female President, Clair Cory. She wanted to run for office at the young, ripe age of 29. She couldn't even get her first executive order done due to the meteors, which was to actually shut down ECTO efforts of sending trash to space.

"Maybe the Colonel can…"

Then it hits me.

Zu! Zu is an Admiral! She has the key to this big ass door. In fact, she *is* the key.

"It's Zu, she's our key. We need to wait for her." I command. As much as I hate her, she's gonna save our asses. "Ijin, contact First Sgt and…"

POW!

A shockwave of a loud, familiar gun could be heard echoing down the field. We look up at the position to where First Sgt's vantage point was held.

As she fired another shot at something.

I run to the Humvee and yank out the radio, breaking the plastic clip that held it.

"*First sgt!* What's happening over there?!"

After a brief no response, I could hear another sound.

One keening with destruction and power.

KA-BOOOOM!!!

The reeking sound of a ring of metal slowly followed after, like a giant sword was being unsheathed. The echo rang all across the air, and the thump of whatever that explosion was felt all the way to us. Dropping the radio, I run out with my gun in hand and look up.

A cloud of smoke and dust covered the area where Ivys, Amaryllis, and Jack once were.

September 27th, 2035

2:36 AM

Houston, Texas

"Contact! Move out! MOVE!!!" I yell. My mind rushing into a silent panic.

Without question, all of the soldiers slid inside their vehicles. I get in mine with Monroe and the doctors, grabbing the radio hanging from the wire.

"Do we have a connection with the Colonel and the rest of the army?!"

"Negative! Only gunshots can be heard from the forced transmission."

"Fuck! Okay, follow me!" I yell, throwing the radio around my shoulder and hitting the gas, screeching and turning towards the direction of the fallen fence.

A low rumble entered my mind when I saw two gray tanks slowly maneuvering their way towards the exact entrance we drove in. Three of them lined behind the fence, waiting for prey.

"Fall back!" I yell, turning and speeding off behind the bunker hill. Fire and steel poured on us as we pushed back. The shots of the tanks are heard raining on us like a horrible, out-of-tune melody. Explosions hit the dirt as we fled. We were only hundreds of feet away. We would be lucky not to get hit.

And so, to speak, one of the Humvees was hit. It shot into the air and slammed almost into a biker. Bullets soon followed, hitting one of our bikers and finishing off that unit.

Six units left.

Finally, with the hill just in front of us, we just made it. To the right of it was a small concrete building of some sort.

"Orders, sir!!!"

So much is happening. This is the real deal. I have to think quickly.

"Take defensive positions! We have a good foothold on the hill. The Humvee with the umm... the machine gun drives behind the small concrete wall to our right. See if you can shoot over it and suppress fire! Everyone else on me! Fall out!"

Monroe and I jump out of the Humvee and try to keep our balance as the tanks empty their shells onto our hill.

"Everyone on me! Get me a bike!"

Yes, I had a plan. A risky, desperate one.

A frontier soldier rides up with one of the motorcycles. I met him before, he was missing two fingers on his hand. He and two of his comrades sitting with him got out and stepped to the side.

"You." I point at the dark-skinned soldier with the missing fingers. "Connect to channel one! And who operates the *grenade launcher?!*" I yell over another explosion.

"We do, sir!" yells a man on my left side. He was one of the tank drivers with Blaze before he died. He pointed over to three frontier soldiers.

"Connected, sir!" the black soldier yelled loudly, cutting my focus to him and the loud radio of crazed laughs.

"YES! YES! FUCK THEM UP! HAHAHA!!!"

That tone. Hurtful and poisonous to my head. The tone I remember long ago was close to being the last thing I'll ever hear in my lifetime. Haunting and animal-like. Just the voice alone sent chills down my spine.

"Don't say anything!" I yell over both the explosions and the radio.

Finally, the .50 Cal M240 was in position, returning fire.

"Sir! Four tanks and footmen hold their ground!"

Fucking great news.

"Alright! The plan! LISTEN IN!" I yell at the top of my voice. "Two Humvees are going to flank the right side! You two-" I yell out, pointing at the grenade gunners. "There is a similar building in shape and size parallel to the one our .50 caliber is firing at now. It's a bad angle gap, so I need you to shoot a smoke screen on it and push behind it along with the rest of the units. When we reach it, you will fire down onto the tanks from the peak above the concrete!"

Funny enough, everyone looked satisfied maybe because it was our only plan because no one else spoke up when they usually do.

"Ijin! Stay behind with me! When we drive forward and charge, you're gonna be shooting down sights out the window! Monroe! Pick six men to flank with you! The rest of you stay close enough to your vehicles and close enough so you can hear me!"

Everyone hesitates for a second.

"WHAT ARE YOU ALL DOING?! WE HAVE A FUCKING BARRAGE OF DEATH HAMMERING US! MOVE!" I yell out, as confident and toned as I could be. Taking all the experience I can from the Colonel's drillings.

"HIT THAT DAMN TURRET THEN?! WHY ARE YOU RATS TELLING ME?!" Trevor's voice ripped from the radio on the bike. **"I'm almost there! I want to see all the carnage!!!"**

I rip the radio microphone from the bike and slam it to my cheek. "Shut your snout and listen up!"

"The fuck?!"

"Listen. and listen well. If you surrender now, I'll consider letting you and your comrades live!"

"HA! Dream on! *BLOW UP THE MOUNTAIN IF YOU HAVE TO BOYS!*"

"Colonel! Sir! Are you there?!" I yell, hoping he's listening in and can reach out.

"He's Dead! All of them are!"

He's lying.

"Why so quiet now?! Crying?!" He laughs.

He's lying.

"Cry all you want! And that little spy truck on the hill is as good as dead! Along with any fuck around it!"

"Your Lying!" I burst out. Fighting tears.

"Look on the hill, and you tell me, kid!"

A tear silently rolled down my cheek as it trembled with my body.

"I'm…"

"WHAT?! SCARED?! GONNA CRY NOW?!"

"I want to kill you."

"I'm sorry. I can't hear the *BITCH* in your voi-"

" I'm going to KILL YOU! WHEN I REACH MY GRASP ON YOU, MY KNIFE WILL *PLUMMET* DOWN *YOUR NECK*. I will have your MONGREL HEAD!!!" I scream out. My tone was new, ripped, and torn from the voice in my head and heart. Both were destroyed at this moment.

"Really now, boy?! Kill them!!! Don't even wait for me. You bilge rats, kill them!!!"

I throw the mic down and threw my M27 on my shoulder, and sprayed the radio and the bike with bullets. My blood felt like it was ripping beneath my skin as I switched mags. Adrenaline spiked through my veins when I racked a new round.

Dominic.

Ivys.

Zu.

Amaryllis.

Jack.

I couldn't protect them. Do they want Irons? The so-called god of revenge was nothing but a title to me. Given by allies and enemies. But now that name will be felt down his ruined bloody body.

"LET GO, MEN!" I bark. My mind was blank on nothing but the death of Trevor Boyten.

The grenade launched a smoke, and we all got inside the Humvees, pushing behind the wall with only bullet dents coming with us.

"WINDOWS OPEN! GUNS OUT!"

The command couldn't be interpreted faster. All the Humvees had soldiers sticking their guns and heads out. Ready to shoot anything that moved.

"Now Ijin! Go!"

The grenade launching its first shots at the tanks was our starting bang to drive and race onto the open field. One of the tanks was already down, while the other was taking grenade fire now. With us charging in, the tanks were not getting a clear shot of us as we sprayed down the men around them.

We charged until we reached another meteor divot. The grenadier moved onto one of the third tanks as it destroyed the second one. Three Humvees flanked their right side and parked behind a street column.

"On foot, men!" I bark into the radio, throwing it down and jumping out.

We all got out armed to the teeth, holding our two new positions strong. Only one man of ours got shot while we took down three more of them. They must not have had a lot of men with them because the remaining five held their ground behind one of the final two tanks.

Where was I? Raining hell down on them with my M27. Shooting down anything that we charged towards on foot. When I ran out of my mag, I popped another. I can barely hear their shots retaliate as the tank catches on fire. We hide behind the original meteor that smashed the fence. Our small foothold. Five of the eleven men followed me. One being Monroe.

"Suppress fire for me. I'm going to smoke the tank out," I command over the constant tacks of bullets.

"Double that advantage," Monroe says, sticking out his arm.

I was hesitant. But if he wants to show me his good side now, then brothers, we shall be. I slam my gloved palm into his wrist in a clutch. "You bitch."

I almost smiled. Turning, "You three! Suppress fire while our comrades run over here!"

"Yes sir!" they yell back.

We held the position until the other five moved over to us. Now was our chance.

"Now, Monroe!"

And we run. Bullets fly around us like hornets. Very angry hornets.

That wanted us more than dead.

September 27th, 2035

3:08 AM

Houston, Texas

We make it to the third destroyed tank with no injuries.

"Irons! Look at this tank! There is a hole in between the tracks. Perhaps these are for ventilation purposes.

These tanks are made cheaply. We can shock them with a grenade."

Yes. The hole. The weak point.

"You go! I'll take out the side cams!" I yell, lying down and aiming.

"Alright," he says as he runs a flank of his own on the tank.

I gotta be quick. The front was already shot out due to Ijins' assault. The tank fired at them as Monroe moved up, not directly hitting a Humvee.

I aim at the back small spinning camera on the gun thirty feet away, and shoot at the hit. Hitting it did nothing.

"*Shit.*" I grilled myself. "Monroe, hurry up!"

The back of the tank opens, and four men pour out—each one with matching gray vests.

I didn't think twice. I shot down at them, hitting two. Monroe raised his gun to the corner to finish them off as they were running back towards me.

"Take cover!!!" he yells in the short distance we had. Jumping down next to me. I slam my chest down next to him.

BOOM!!!

The top hull of the tank flew up into the air like a popped soda, slowly falling back down with a heavy thump.

I pick myself onto my knee as Monroe stands up and offers an arm. I take it and dust off my face and gun.

"Your hair looks like shit."

"Fuck off," I respond.

It was quiet. The bullets all stopped, and the only sound was the whooshes of the fires burning around.

We won. And lost.

3:30

We met back at the entrance of the bunker again. The headcount before was 38. Now, only 27 remained. We had four Humvees and one shot-up bike. (Oops.) It still works, though.

"Sir, our ammunition is depleted significantly," the black frontier soldier from before says to me.

"What's your name?"

"Finn Ricco, sir."

"Finn. Okay, thank you for informing me and all your help," I tell him as I begin walking past him to my Humvee. I get inside and grab the broken Radio mic.

"Will?" Adonis says from the back.

I slam my thumb on the channel two button and bring it to my mouth.

"Where are you?"

`"You think you scare me, kid? I killed hundreds of you. I pillaged more towns than you can name. Your school base is dust and ashes. You're `*`nothing`*` to me."`

He's trying to get to me.

"Too much of a *bitch* to show yourself?"

`"L#- Lutenet! Do you read?!"` The voice crackles in static over the radio.

It was Colonel Dominic's voice.

"Copy, sir! We are at the base! It's real! Over the turnpike on Aldine!"

`"Copy that. Hang tight, lieutenant."`

`"You really think you're gonna reach Iron's body `*`alive?!"`* Trevor laughed.

`"Find cover, Will!"`

"Will?" Trevor said, puzzled. `"`*`WILL LIGHT?!`*` YOUR THAT `*`BRAT`*` I DIDN'T GET TO KILL BECAUSE OF THESE `*`FUCKERS`*`!"`

I could almost *hear* him smiling.

 "You think you're a soldier boy? Do you think you're the revenge God? COME SHOW ME THEN!"

I drop the radio and run out. "Get behind the bunker again! *Now!*"

 No one questioned me. Everyone ran to their vehicles and started speeding around the mountain. Monroe and I were the last ones to follow when I see it. It was emerging around the corner was one tank. Cruising anything in its way.

It was huge —huge enough to crush the other tanks. The sheer size of this tank was bigger than a commercial plane. It had a six-barrel gun with two machine guns mounted on the front. What the fuck is that thing. This thing was painted fully gray, aside from the messy black spray paint on the hull that read "God eater."

We made it behind the hill before the barrels could take a shot at us. The 50. caliber drove behind the concrete building again. Ready to take fire.

I switched quickly back to our channel yelling "Don't! We don't know if-"

Before I could finish, the tank fired at the concrete box, blowing it to smithereens. I felt the damage before I saw it. The shot blasts our eardrums and flattens the Humvee to the ground.

"F… Fall back!" I yell.

"Sir! The guns will hit us before we reach far enough out!"

Behind us was a long bush forest with a seemingly open field behind it. Aside from the rubble, that is. But that gun could probably shoot well over that tree line.

That screeching noise of bending metal came after again. It lasted for eight seconds before it silenced for another twenty.

It fired again after, practically putting a dent in the concrete of the bunker. The thump of the wave almost knocked us to our knees.

Giving me a plan.

"Everyone listen up! Squish inside these three Humvees! I do not care *how,* just do it! Now!"

I run to the backseat of my Humvee, where the doctors were. I started ripping the door open. "Out now! Get in any other one!" I bark, causing them to hurriedly step out.

"We removed the backseats so we can all fit, sir!" Ijin yelled.

"Good!"

I gotta be quick. The barrels cranking ended, I had twenty seconds.

I climb the grenade launcher and grab two rounds: one smoke and one live.

"Monroe! Get that bike! Ijin! On my signal, I want you to drive for those trees like you stole something! *NOW!*"

He runs and gets inside one of the Humvees, while Monroe drives up in the shot-up motorcycle. I grab a heavy piece of smashed concrete and walk it to the Humvee.

"What are you doing?!" Monroe yelled loudly.

I make it to my empty, reddened Humvee and throw down the boulder on the gas, jumping out when the car jerks and drives off to the right.

Three seconds later...

BOOM!

Destroyed in an instant.

I shot my M27 in the air afterward. *"MOVE!!!"*

It was as if I hit a horse's thigh.

They all gallop towards the trees as the super tank reloads.

"Do we run with the bike?!" Monroe questioned, kicking off its kickstand.

The turrets of the tank tore down at the running fleet.

"Hell no! We will be shot down! Listen. That thing is gradually getting closer. I'm going to climb it and infiltrate it when it reaches us.

"You wanna die or something?!"

"Do you wanna live or something?!" I yell, watching the Humvees make it past the trees.

I peek over the hill as the thousand-ton monster hurls itself at us slowly. But for its size, it's fast.

"Lay down now. I'll tell you the plan," I say.

He listens, while shrouding himself under the hill as another round shoots blindly through the trees.

"Good. They escaped."

"The fucking plan, please?!" he hisses.

"You're gonna boost me up on the tank when it pulls around by us. When you do, you're gonna take this-" I hand him the smoke grenade "-And smoke out the outside cameras. Take that bike after and run for it. Push for the street columns until you're safe."

"And *yourself?*"

I switch my mag and reload my gun, taking off my coat and tying it to each end of the gun, making a robust sling that I threw over my shoulder. "Gonna infiltrate the tank, of course," I say as I drop my bag and pull out the black flare gun, stuffing it inside one of my left pockets, in case of anything.

"This and a stupid ass idea. Get ready," he whispers. "And get your ass back."

No shit, I wanted to say.

The tank's tracks start pulling their way past us, ripping up the dirt and ground.

"Now," I whisper.

We get up from our crevice and run to the mammoth's side. We noticed a metal ladder eight feet high as we ran beside it, avoiding any cameras.

Monroe ran forward and turned quickly, interlocking his fingers. It allowed me to step up to the hull and reach the ladder.

"Don't die," he warned. Lighting the smoke and running to the bike, he drops it down and rips dirt as bullets whizz past him.

No turning back now.

I make it up the ladder and see the hatch of the entrance inside. I pull the pin of the grenade and stick it under a pinch of the door. I continued running back behind the ladder until I heard the explosion and saw the wind take most of the smoke. The hatch popped open with ease after.

I threw my gun back at rest with my shoulder, jumping down the dark hall. The walls around were armored, and the lights inside blared red. I could hear the yelling and the cursing inside.

With my gun raised, I ran across the thin hall. When I reach a split off of three halls, I start with the left one. When I ran down the hall, I could see men.

Waiting.

I saw one with his legs sticking out as the hall made a sharp right turn. I shoot it and run up to him as he jerks forward, grabbing his vest and turning the corner with him in front of me. He's yelling in Spanish while slamming his fist on my bag. The two Sentinels behind him start shooting him in an attempt to shoot me down. He covers me as a human shield. As they run out of ammo, I drop him and lace both of their chests with bullets. I ran past them through the hall and into a small technological room, it was completely dark aside from the flashing red lights in each corner. In the center of this small room was a Sentinel sitting on a gray leather swivel chair. Screens in front of him showed the Turret POV. He soon heard me and grabbed a fire extinguisher that was next to him and began to get up. I shot him down before he could stand, emptying my mag left onto the screens of the turrets.

One down. Two to go.

I go back down the hallway and grab a grenade from one of the Sentinels' bodies. Running down to the farthest left hallway, I threw it around the corner.

The explosion catches them badly. Even if it did just stun them, I turned the corner and sprayed the rest of the dying men down. Blood trickles down the walls, staining the roof and my pants with a thick, heavy ooze of black and red. I ignore it as I run down the hall.

Suddenly, my gun is pulled away from me, and I'm shot in my left shoulder with a handgun. The pain wasn't instant, but it came after I grabbed it and wrestled him with it against the wall. It didn't feel as bad compared to the first time I was shot. Like I got poked heavily in a pressure point. I have no idea if adrenaline took the sting.

The pain only sharpened as I wrestled him. I ignored it as I slammed his hand onto the wall, making him drop his gun, giving me the opportunity to slam his head next. He stumbled and fell face-first to the floor, giving me the opportunity again. I jumped on top of him and grabbed his hair. Slamming his face on the floor,

Over

And over

And over

Blood spews from every orifice of his face as I continue. Splashing and warming my body and fueling the rage that flowed down my spine. I dropped his body, leaked down the floor everywhere, and shot out a similar pair of screens that were from the other side.

Fuck, it was hard to aim. I could barely shoot with my newly fucked up arm. I strap it back to my back and tighten the loop to the coat that slung it to my back. Pulling out my pistol, I stumble down the hallway and back to the split-off. The lights seemed darker, and the sounds around were replaced by my heavy heartbeat that strummed with every movement. It was louder than my thoughts. It replaced the focus on my mind. I scrambled myself into the final hall and reached a similar hatch on the outside. Only this one was open.

I jump down with a slight sting.

"Good to see you Will," Trevor said from in front of me.

September 27th, 2035

3:42 PM

The God Eater, Texas

The sight was *horrid*. A close-to-literal blood bath was inside. Soldiers' bodies were ripped up and thrown in a corner. The arms and legs were stacked over the men they once attached to, thrown loosely in another corner, bleeding all over the men that were below them.

The room was small. Though it was futuristic-looking, a white dystopian painted an ugly picture compared to the red gore inside. On the left was a large boiler, and in the center was a large black chair with a big clear screen in front of it, similar to a window.

Similar to the other screens of the turrets.

"Thought they would get in the way of our first meeting, Mr. Light. You know how important first impressions are. Though this is our second-"

I shot the screen behind him before he could finish.

"God damn it!" he yells, cranking a lever down that caused the room to jerk. "There! I just stopped it, so our talk is legible!"

"I'm not gonna shoot you."

"*I know,*" he snarls.

Fuck. This guy is *toned*.

Veins sprung from arms up to his neck. He was bare-chested and bulky. Tribal tattoos covered his abs and chest in thick black and dark green ink. A 6'5 gorilla of a man, ripped cargo pants, black boots, and a weird sprawling silver pendant hung from his neck was the only set of clothing he wore. His hair was wild and long, his eyes purple and dangerous.

And his chest. It had a double-sided war ax pattern, similar to the one the nurse had on her face.

The pain she must have felt.

"Boy. I am your king. Everything *kneels* under me. I am beyond strength," he growls, wiping blood from his cheek.

I could feel the blood trickle down my arm as I holster my gun.

"HAHAHA!" He grips his head before charging at me and punching my jaw. I could feel the blood leaking down my nose instantly.

I manage to raise my hands for the next hit.

Next came a kick to my side. I managed to grab it before he pulled back and slammed my elbow onto his knee.

Holy fuck! He was built of concrete. I put so much force in that hit that it would have completely broken a normal man's kneecap. All he did was stumble back. I didn't waste a second… not again. I ended my move with a palm strike to the throat.

"*GRRK! YOU LITTLE SHIT!!!*" He chokes as he charges back at me with a meaty fist to my stomach.

I heard the snap inside my… inside me. It pulsed my next cough to spit out an ounce of blood. I stumble back up and away from him.

"*Too bad you won't be seeing that pretty girl anymore.*"

At that moment, something spiked inside me. Not the cracked bone or the exposed wound. Something new. I throw down my bag and my M27. And charge my full body at him in a tackle. I slam my fist into his face three times before he grabs my hair and throws me away from him. I scrambled back to my feet, grabbing the black blade on my hilt. Charging at him

Jumping on his back as he crouched over and plummeted the knife into his spine. His cry of agony shot my eardrums as he grabbed me from his back by my right arm and flipped me over him. Slamming me on it.

SNAP!

"*FUCK!*" I cried before getting on my knees quickly and charging at him again. While I pulled my arm with me as it dangled loosely, he was pulling the blade from his back. I kick him in his nose, and he slumps onto his stomach. Not knocked out. Just very, very angry.

I pull the blade from its beast stone and try to stab the back of his neck before he flips me off him, and he flips to his back. He threw my hand away again and again, but it didn't matter. I plummet the knife close enough to his neck to where it sliced close to deep enough. His blood grew darker, and his strength weakened as the veins in his arms poked out vigorously.

"My face will be the LAST THING YOU EVER SEE!" He shouted, waving my pistol in his hand for me to see. But he wasn't trying to aim at me.

He pointed it at the large container of bubbling yellow liquid.

Before I slammed the black blade deeper into his neck, he shot the boiler. The glass shattered, washing all the blood that leaked out of his neck. His purple eyes dilate as the heat of the room rises to stinging temperatures.

September 27th, 2035

14:30

Houston, Texas

I missed my Will.

So I complained to Ivys.

"Ivys…?"

"Yes. I know you miss Will. It hasn't even been a minute since you told me last time," she scoffs. Not moving an inch from her position.

I sigh as I lay down next to her. "But I-"

"Look, he hasn't even been gone twenty minutes. You can see him from here. He is fine, Amaryllis.

Our position was held on a hill. The high view allowed me to see Will exactly two miles southwest-ish for me. I could see him step out of his car with his friends and walk up to the evacuation unit.

"I don't think they are going to be able to get inside. They need a person with class 5 government clearance. And they are probably inside the Pentagon or the Whitehouse."

"Look… can I just-" Ivys begins, stuttering. "I need to ask, I get that you don't remember anything, but what happens when you remember like… *that?*" she asks as she turns around and lays on her back, taking a swig from her Canteen.

"I don't know. It's just like thoughts… like certain words trigger responses. A ring of a bell triggers a boxer's thoughts. A beep of a machine triggers a marksman's thoughts. It's like… developed in my mind.

Ivys stares as if I spoke another language.

"I'm sorry," I say quickly, avoiding the harsh eye contact.

"You probably are, if not more curious than I am to this. I understand. We have more-"

I completely blacked out on her next words. I poked my head up and saw… *it.*

"Ivys! Look!" I yell, pointing down to a…

Um.

"What the fuck *is* that thing?" She whispered, dropping her canteen and looking through her scope that pointed directly to a giant rolling piece of iron that had a six-barrel gun on the top, treading easily through the ruined landscape and ruining it even more.

Without question, she shot at it. The loud *ting* of the graze heavily dented the beast.

I heard the truck knocking as we inspected the tank. Jack's muffled voice leaked out from the tough walls of the Kraz 6322.

"*Amaryllis! The radio!*" I could hear him yell faintly.

I run to the driver's side and open the door to the loud chattering of booms and cracks.

Ivys snuck from behind me and grabbed the radio transmitter. "Give me a tactical Colonel!"

"They are hammering hard, but we have the field advantage! We are continuing to pull forw-"

The transmission ended with a crash before he could finish. The only thing that followed after was a low Buzz.

"Colonel?! Colonel, do you read?!" I hear her yell, fumbling with the radio

My eyes focused on the large tank. It was not so far away from where we were, but it now pointed its six barrels directly at us. I should know because I could see directly down them.

"Ivys!" I could hear her giving our location before I pulled her out of the truck and to the floor.

Half a second later, a loud Bang closed off. As if two Anvils were thrown against each other at full speed and collided directly.

With us in between them.

The impact hit the truck side, and that's all I was able to see before the solid kick-up of smoke and dirt. The crash rang my eardrums intensely for about ten seconds before the muffled ambiance came. The only thing I could see was the dust and smoke floating around in a thick fog.

"Jack!" I scramble to my feet. Stars fill my vision while I try to call out his name. My heart grew louder and harsher beats in my chest the first couple of callouts."

"I'm okay!"

As the fog cleared, I could spot the side of the truck completely ripped and destroyed. Inside was Jack. The only difference was Vines covered his body thickly weaving around him head to toe, shielding him.

I cough out the dust of the loud gasp I gave afterward.

"*Damn those sentinels!*" Ivys yells from the clearing smoke. She was okay. She clutched onto her sniper with one hand while she tried to cover the oozing blood on her leg with the other.

I marched through the rubble to her and inspected the massive slice on her shin. It reached about six inches and cut at least an inch or two deep.

"You're injured badly," I quivered.

"Shrapnel. I'll be fine," she says, using her sniper for support as she struggles to get up. She tries to walk without the obvious limp in her step. It was of no good as it was causing more blood to leak down her boot. I soothe her as she sits down, using some very friendly Pirate tongue.

"Can I borrow your knife?" I ask, eyeing the one on the hilt.

She silently pulls it from behind her and hands me a six-inch bayonet blade. I begin to slice away the parts of her pants around her cut. Jack finally managed to free himself from his little green cocoon and walked over.

"So… this is the boy?" Ivys asks, looking at Jack's flora-filled hair.

"Yes… Hi." He softly answers.

"Ivys. Skipping introductions, whatever the fuck that was, it destroyed our trauma kit and our radio."

"It was a T80 Mammoth tank. A class 11 tank developed in the early 2000s. Left abandoned as a prototype." I blurt out. Confused on where that came from.

The air around started to fill with distant sounds of warfare.

Daffodils, I hope Will does not have to deal with that.

Ivys just looks at the ground and chuckles as I finish removing the little pebbles all around her cut.

"Jack, I need two firm sticks or poles. Something that can hold together the whole bottom half of her led as long as possible. Can you find them for me?"

"I'll try my best," he says, running off into the rubble, dragging vines behind him.

"What the hell *is* he? Are those… *flowers?*" Ivys harshly whispers, lying back.

"What he *is* doesn't change who he is, which is a *normal kid*. He just has an… *ability*. He can grow flowers from his body."

She stares at me, and a mix of awe and disbelief crosses her face. She sits back up and looks at him again as he digs through the rubble.

"He really does look like Will." she smiled slightly.

I don't know why, but that reminded me of Silvia, of who I could truly be.

"What is it?" She asks, wincing as I dab her cut with the shirt strip that was tied on my waist.

"I think I found out who I was. Who I really am-"

"Stop." She raises her palm. "Who cares? You're Amaryllis."

I just silently nod to that, arguing no further. She was right in ways. Whoever I was… whoever *Laura* was is gone. And I need to remember them.

"Amaryllis!" I hear Jack yell while running over with two straightened branches in his hand. It was possibly blown off from the dead tree when the tank fired.

I line the branches on either side of her leg, making a robust framework for a stilt, using the piece of cloth to tie one end tight.

"I don't have the supplies to help much…" I silently said.

Without a word, Jack placed his hand on Ivys' leg. The short vine from under his arm started to grow intensely fast as it wrapped around Ivys' legs in kindle twines and twirls. He begins to tighten them as flowers bloom off the top of it.

"Is it comfortable?" Jack asks Ivys with a slight grunt.

"Y… yes," she answered, dumbstruck and awed.

"Amaryllis? Can you cut me free?" Jack asks, shaking his vine-tied arm.

"Mhm." Silently, I slice the first vine, making Jack wince while he pulls away.

"Thank you, Amaryllis. Can you… cut the other ones? It kinda hurts when they snap, being pulled from rocks and whatever…"

I get up and start cutting the wings of vines growing too long and loosely as he poses in a T pose. He was sweating heavily while red when I continued.

"Hey, Jack, was it?" Ivys asks, not lifting her eyes from her leg. "Thank you."

"Mhm. Ow," he complains softly as I finish up, cutting off one final vine on the back of his leg.

"That's the best I can do," I say softly.

"If my transmission got sent, the Colonel should be here any minute," Ivys informed, pulling out a watch from her pocket.

I failed to clear up my surroundings.

I was heavily reminded as I heard another loud bang from another destroyed target.

"Did it reach Will's position that fast?!"

"We need to worry for ourselves," Ivys called out. She puts away the watch and sits up. "All we can do for them now is hope. Hope kept us this far. It will pull Will farther."

About a minute on cue, a Humvee tagged in red pulls down the street and into our divot of problems. It stops a few feet from the rubble and Dominic steps out the driver's side.

"Ivys?"

"Yes, sir," she answers, getting up and trying to avoid her obvious limp. I put her arm around mine and walked her over.

"Where's the troop?" Ivys asks. She was clearly not concerned over her dysfunctional leg.

"Easy, Sgt. You need to be concerned over yourself. The troop is fine, and they are heading over here when the ambush is finished. All we lost was our prisoners of war."

"Sylvia?"

"Negative. She did run, but she took her own route." Dominic looks at Ivys' leg and furrows his eyebrows at the flora-wrapped wound.

"Sgt..?"

"The boy's fine work." Ivys jabbed her thumb at Jack, who just tried gaining a straight posture

"Uhuh."

"It's a long story-" I begin "-but in short. Jack here, he can manipulate his flowers and vines at will."

Jack then picks a flower from his neck and walks up to Colonel, holding the moody violet plant out for him to take.

"So… Will is your brother?" He questioned, taking off his glove and taking the flower in his hand strangely covered in burn scars. All I can tell is that it healed long ago. "You both do look similar… aside from the plants," he says, sticking the flower in a pocket on his shoulder.

"Get in. We are going to that tank, I made contact with them on the way here."

"You're bringing the kid?!" Ivys yelled.

"We can't let anyone discover him yet. He was kept a secret for a reason, so we are just going to hold him."

"It's an active *battlefield!*" Ivys again yelled. I could see her face crossed with nameless emotions.

"He's going to be safe inside. Amaryllis and um… Jack." Dominic begins as he walks back to his Humvee. "Hop inside," he commanded, stepping inside and starting the engine, warming it up for its new passengers. Us.

"Thank you for protecting us." Jack turns to Ivys, hugging her quickly and running over to the Humvee.

And Ivys, after her surprised look, she turns to me.

"Protect him. *That's an order.*"

"Yes, ma'am." I nod. How could I ever disobey this order?

14:51

We drove down the high point in silence. My fingers tapping the windshield, looking out. I'm hoping to see Will running over any minute now…

"Excuse me, sir?" Jack's tiny voice came from the back seat.

"Yeah," Dominic answers, still tough-sounding but… softer than usual.

"What's your name?"

"Colonel Dominic of the first Rebel Independent Ruins Army."

"Ah…" is all Jack responds with.

"Just call me Dominic," he responded quickly

"Thank you, Dominic."

"Of course. Now, understand where we are going. You could get hurt if you don't stay inside the truck. Or worse. Got it?"

"Yes, sir," Jack responded.

"How are you controlling the Vines?" I asked, hoping there was not a flower forest back there.

"It's good, though the feeling is weird. It's like… manually blinking.

"The tank came to a halt. Maybe something jammed," Dominic concludes quickly. "Amaryllis, watch Jack," he commanded. And without warning, he stepped out.

I didn't even get the chance to say goodbye. I was fixated on the 30-meter ton of steel to the right of us. A monster of war that can turn any battlefield into dust if used correctly. The only thing out of the ordinary was Dominic climbing the ladder and entering the already open hatch at the top, his gun drawn.

"I feel something, Amaryllis. It's… Will," Jack whispers, "He's inside."

"It's okay, Dominic is going to-"

My words slowly fall from my tongue as the rest sink back into the back of my throat when I see Dominic rushing out. No Will in sight. He climbs down and rushes inside the Humvee.

"Dominic?

"What-"
"We gotta move. That tank is gonna blow itself up. Hitting anything around it with whatever comes out, I saw only smoke inside, and it was very hot. The boiler somehow burst, so it's a ticking time bomb. Best to be at a far range."

"Wait! Will is inside! We have to pull him out!" Jack started shouting.

"Will?" Dominic turns.

"He's inside! Please get him!" Jack begs, sitting up as the flowers in his hair begin to bloom more.

"Look, Jack, even if he is, the smoke will make it impossible to see him more or less, if the heat doesn't knock me out first. The tank can explode any minute now," he explained, pulling the stick down into reverse.

"NO!" Jack howled as he opened the door and jumped out.

September 27th, 2035

14:58

Houston, Texas

"*Shit! Stop him!*" Dominic yells, slamming the brakes.

He then tries to get out.

Tries.

"The hell?!" he yells, slamming his shoulder into the door.

I try to do the same. With not so much of a budge, I only managed to spot a green rope from the bottom of the door.

"It's Jack! He locked us inside."

I could see him cutting the vines connecting him with the Humvee. He used the same Mayo Scissors he used to cut me free the first time. After slicing the final rope, he runs over to the tank hull and places his hand on it. Holding it there.

"*Jack!*" I croak.

Dominic continued to slam his full body weight onto the door. Finally, the door gives in and opens a crack—vines catching between it. Dominic pulls out a knife and slices the flowery vines, jumping out.

KA-BOOOOOM

The pulse shook the ground, and the top half of the tank popped. An explosion of smoke escaping like fireworks came from each motor of each joint of the machine. Each windblast pushed Jack over and over until Dominic reached him and shielded Jack.

And Jack started growing vines around both of them. Anchoring them to the ground.

And oh God…

Will.

All that I saw from my view was smoke. Black thick smoke endlessly steamed out. It reminded me of a large ghost.

Oh, Will.

My head started to pound again. I see Dominic cut himself free from Jack's shackles of plants. Dominic runs over to my door side and cuts the vines, opening the door. I couldn't let him see me or anyone else. I kept looking at the floor, hoping my hair covered my face. I needed to hide my burning emotion quickly. Dry up. Jack needs a clutch more than I do. I look up and see him standing in the empty field, facing away towards the destroyed tank. The top was ripped and smoking, similar to a can of burnt food.

I got out slowly and sulked myself to Jack.

Who wasn't sad?

"Are you… *Smiling?*"

"Will is okay! Come!" He yells, running to the Mammoths later, and begins climbing up the 18-20 ft. climb.

"What are you talking about?" I call out to him, following behind as Dominic follows behind me.

"Jack! There is no way he could survive this!" Dominic called up to Jack as he turned back to the Humvee.

"Jack, it's okay," I say. Finishing my climb soon after.

Only to leave me with a light gasp.

At the top, the inside was exposed. Barely anything shown from the original parts of the walls of the interior. Why? Vines and flowers covered it all. Everything looks like it hasn't been touched in decades. The smoke only came from the cooling of fires that the flowers covered, still making things hard to see.

"It worked! Look!"

In the middle of the destruction was a clustered pile of darkened vines and burned flowers. Charred and recognizable. Inside the center of that cluster barely sticking out the top, was a lock of silvery blond messy hair.

"Dominic! Come quickly!!!" I basically cried out. He was busy pulling off the vines around the wheels before he rushed over. Climbing the ladder up and marching towards us. I point at the cocoon of vines with Will inside.

He jumps down and cuts the vines around Will's toughened body. More and more of Will's body is shown as Dominic continues to cut through the roots and vines. After his head was mostly shown, Dominic ungloved his hand and pressed two fingers into the side of his neck.

"Well, ain't that a bitch," Colonel Scoffed. "He's alive."

At that moment, my heart felt full again. I couldn't contain the tear that rolled down my cheek.

That then turned into a stream of tears.

A good cry took over my emotions.

Jack managed to make it down to Will with the Colonel and grabbed a bag covered in chars and blood —Will's combat bag. The one I read was from his father.

So I climbed down slowly to them. Getting closer to Will's body, I could notice that he wasn't exactly the cleanest either. From his cheeks to his arms had extensive blood. Too deep for the cuts he had that we could see.

Jack digs through Will's bag, pulls out a small white MedKit, and hands it to Dominic. And shortly after Dominic was able to cut Will free.

And when I tried walking up to both of them, I tripped over… a body. Which gave me a mini heart attack as I crawled away, panicking. The man was large. Very large. His bulky body was stained with blood, cuts, and bruises. His long hair was burnt and messy, almost hiding the thick slice on his neck. The only thing pretty on him was his eyes gazing wide open.

He was dead.

The purple gaze was gone and dilated from the death that took him.

"*Trevor,*" Dominic says quietly. "The Beast is dead."

He goes back to the MedKit and pulls out two white sticks. I instantly recognize the pungent smell of the sniffing salt embedded in them. He takes both of them and snaps them right under Will's nose.

"*GAH!*" He jerks up and yells, throwing a solid left hook at Dominic's cheek.

"Will! It's okay! It's us! We won!"

He stops for a second and inspects me before groaning loudly and laying back down in his vine bed. Dominic had already recovered, rubbing his cheek.

"Amaryllis…"

"Yes! And Jack! He's here too!" I practically croaked with excitement.

"Am I dead..?"

"What?! No! Of course not!"

"I feel like death. Is my mom here?"
"Will!" I scream, clutching both sides of his head and… uhm…

Okay, I don't know what I was thinking! I shove my lips to his. His passive response was only to allow it. Tears streamed down my face, resting on his as I twisted my lips around his. He closed his eyes and let go of any resistance, I could taste the blood on his tongue.

Then I slowly pull away, trembling slightly and hoping not to forget everything and get lost… again.

"Oh my God you're all alive!" He snaps. "You're alive! Jack!" He calls, trying to get up. But then he immediately crumples back down. He sucked his teeth as tears came now from him—uncontrollable watery eyes.

"Woah. This… *hurts.* I hurt. I can't move…" he quivers. Trembling and sweating

Okay! Okay! Relax, lieutenant, the fleet is coming back now," Dominic soothes."

Then I look at Jack. He stared back, still surprised by my unexpected… show.

"How?" I asked. Or more like demanded with my croaked tone.

"A long time ago, I gave my mom a butt of a flower. She put it inside a necklace and kept it at home. I guess she gave it to Will because I felt the force… the life energy in the flower. So I covered him."

All I could do was nod because I was speechless.

"Thank you," Will whispered. "But shouldn't you be hiding?"

"I'll handle whatever comes my way. No more secrets," Jack nods, picking up a black spray-painted flare from the floor.

"Glad I didn't get to use that thing," Will chokes, grinning slightly.

Jack pulls the top of the flare and holds it up. The army all rounding up, moving past the bunker and pulling in as he fired the red glare.

September 27th, 2035

3:31 AM

Evacuation Unit 12. Houston, Texas

The next hour was the same feeling for me. Painful.

Actually, *agony* is a better term.

My diagnosis from the doctor came back. And the medical bill was heavy. A gunshot wound nicking my scapula. Two ribs were fully broken, fractured hands, fractured right humerus, and finally, a broken left arm — not to mention the bruising all over my body, the severe cuts and scratches on my skin. I didn't mention it, but my scalp was hurting like a bitch too. My eyes burned enough looking around on their own. They felt worse looking at me.

Other than that, I was in *peak* condition.

I was sitting up in a Humvee alone, an IV sprawling around me, similar to Jack's vines. Bandages covered me like blooming flowers. All of my clothes and gear were bloodstained. My bag was a little burnt, but it was still usable. My gun strapped to my back caused severe burning on it, I wouldn't be surprised if it left its own brand mark itself. It rested next to me, and finally, the black combat knife in my shattered hand slept peacefully there ever since Colonel gave it back to me. Thanking me while doing so.

"For what?"

"With Trevor dead, the world is ten times safer. You're giving us all hope."

"Hope," I repeat, sighing

Hope was what kept us all alive. The only thing stronger than fear.

I felt like I read that somewhere…

But not everything was getting better. The truth was that a fleet of Sentinels pursued our men back to our home camp.

There were no survivors reported.

I don't know how I'm gonna be able to tell Jack that Alice is gone. That *mom* is gone. It was hard to swallow the information myself. In fact, I'm still choking on it.

The last thing my mother left me with was the thing that saved my life, leaving her unprotected.

Now I have nothing from her.

All I left for her was a stupid gun. I couldn't be there for her. I can only look at the limited hopes I had.

And that was Jack.

I fight to tell myself that it's all mom wanted. And I lose that fight at the part where I had to believe it.

Suddenly, the door opened. Ivys poked her head in. "Can you walk?"

"Yeah," I start pulling out needles and tubes, skipping the part where my entire persona was wincing at the very definition of "Pain."

Ivys grabs my wrist softly, pulling me up and carrying me out, holding me up under my arm. She hands me my M27 to use as a crutch.

"You should be as good as done for the next month," she sighs.

"Great. While our enemy grows in number?"

"You're just a kid. You shouldn't be worried about anything related to war."

"As if you tell me that when you made me a *lieutenant*." I scoff. Clutching on my cane of a gun

Her reaction doesn't change, but I can feel the energy of a shock hit her. Why? I was shocked, too.

So I changed the subject.

"How many of us?"

"A little over a hundred. Not including civilians."

The walk goes silent as we make it to the door again, now artistically scattered with new rubble. The corner of the wall is clearly blown all over the path. Luckily, the hand scanner isn't destroyed. In fact, it was not even too tarnished—a lucky break for us.

First Sgt brings me to the Colonel, who started talking before I could even ask.

"No. I'm not allowed in either. Guess the superiors thought I wasn't necessary for this kinda thing. Gave me a whole "Access denied" speech."

"What about Zu?"

"Access denied."

"*What?!* She is an Admiral! I thought class five could access the bunker."

"Yeah, we know. We tried Ivys just for the hell of it and nothing. Must be private property. That's the only solution that we-"

"*ACCESS GRANTED!*" Shouted a loud, monotone voice.

The door started making piston sounds of releasing air and slowly started to open inwards. Everyone raised their guns.

The only one standing at the Scanner, her palm perfectly placed, was Amaryllis. Keeping her hand on the green glowing screen as she looked back at us.

7:20 PM.

"*Guess she might be the President's secret daughter or something.*" First Sgt chucked as I moved into my dusty cot right by the stairway exit—11th floor down.

"Whoever she was, she is now Amaryllis."

When the door of the bunker first opened, Colonel quickly formed three teams to raid inside the whole bunker. It took a while, but we counted twenty floors. Five thousand or so square feet shown all around each floor. God knows what it could look like with all the lights on. It was incredible, also it had no people inside, the first ten floors were just linings of bunk beds. The other ten consisted of two cafeterias, two medical offices in which one was locked off, a meeting stage, a kitchen, and two vehicle storage bases at the top, both reaching a size large enough to be considered four floors.

Clearing a path inside was fine. However, there was just one teeny tiny problem.

"Power generators were knocked out during the skirmish. We will have it back in a day or two."

Great. Now, all the supplies that are supposed to fix me are all locked away until the power is on.

I can't complain, though. We are safe.

"I gotta go." First Sgt broke the ice. "Rest easy. If you need anything I'll be helping the crew trying to break into the fifth-floor door."

"Medical?"

"Agricultural, actually. The one we thought was Medical was actually a small self-refurbishing farm. That room alone could feed a thousand of us if used correctly."

"Then go, I'm starving." I groan, laying down in my bottom bunk as all my broken bones lay to rest. And God, I am *not* going to be able to sleep tonight.

I feel something thump on my stomach, causing me a little pain. I was about to yell at First Sgt until I saw it was a protein bar.

"Can you give it to Jack instead?" I ask, not moving.

"He is still with the doctors. I'm sure they have already fed him. He will be fine," she says before turning and walking off into the darkness.

Actually, more like dimness. Almost all the bunks had some sort of light source, I was lucky with my state-of-the-art lightbulb connected to a double AA battery. I also was lucky because I was given pain relief, so I wasn't in complete disarray. But I still felt like I got jumped by a crazed animal ironically.

What's a little new to me was that they didn't split off females from males. My guess is that in the situation we are in it is hard to get distracted with lust. Everyone here is mature, toughened, and miserable. I wish that calmed down my overthinking mind. Because, at this point, nothing could.

"*Will?*" I hear a tiny angelic whisper call out.

"Amaryllis!" I shot up from my bed.

Which was a terrible, terrible idea.

"It's okay! Relax…" She says, sitting on the side of the bed. I ignored the tears that came with the pain and sat beside her. Stars fill my vision.

"I heard the news. From what I know about your mom, she seemed so lovely. I'm so sorry, Will."

"It's fine." I sigh. I don't know how many of these lies I can keep juggling.

"And… about yesterday-"

"It's okay." I quickly say. Looking down,

She leans over on me and takes me in a hug, and I just sat there, stiff. "Amaryllis, it's okay…"

"It's not that Will…" she says, burrowing into my shoulder. "I'm just scared. Emotions are confusing."

"I feel that," I answer, leaning on her. "It will sort out on its own. We just need to be patient and wait."

"Breath…"

Oh God, now she is trying to comfort me. She thinks I'm in distress when she is in *disarray*. She is the one suffering in silence way more than I will. God forbid she remembered her *life*.

"I'm okay. I'm so sorry."

"No..! breath..! I can't..!" She wheezes, turning blue.

"Oh!" I got off her when I realized I was crushing her stomach to the bunk. "I'm so sorry!" She giggles and pets my head, gently weaving her fingers through my scalp and twisting around my hair strands. It hurt like a bitch.

It was also the best feeling in the world. It felt like sunlight. Maybe she saw my shocked, stricken face because she kissed my cheek just to make it *worse*. "We will be okay, right?"

Suddenly, the whole room lit up. The lights… power were back online. Happy soldiers' hoots filled the room.

"Hope, Amaryllis. One who falls too deep will see a dim light too bright."

September 28th, 2035

0800 hours.

King's revenge. Gulf of Mexico.

"My my, I'm appalled you made it back," I could hear him say. Dressed in a large white lab coat, and chuckles a cold and soothing tone as he stares down Silvia. Between them was a solid dark oak desk, and a dimple pencil, paper, and gun were laid out at the top. Sat resting in a room of silver and black walls.

"So how did you get here?"

"I escaped during Beast's ambush. Most of the sentinels led sacrifices so I can be here."

"Most." He cuts off in an annoyed tone. "Most is not all." He then grabs the 1911 and cocks the handle back. His straight silvery hair receding and his face pure and pale, that of a ghost. "This aircraft carrier alone has a hundred men. We have two of them, along with extra battleships and a submersible. Our trip to Florida will be straight, with *nothing* getting in our way."

"Sir, what about the others? There is far more than-"

"All require codes, don't change the subject," he commanded, walking around the desk and pointing the barrel of the gun at Silvia's head. His 6'2 frame towering over her.

"Warum sollte ich dich nicht töten?"

"Father, I knew death followed if I ever returned with betrayal. I came back to promise you the Colonel's head of the Red Army."

"You have failed already! What puts my trust where I believe you won't fail me again, *mädchen,"* he says, stiff and calm.

"Omen sir, to die by your hands would be an honor, but to serve you with every ounce of my life would mean my entire gratitude. I show only my deepest respect for you, sir," Silvia says, looking to the floor.

Omen sighs in response, "I have all these AI drones and droids to control my units. And yet, I do need someone reliable on the inside. The experiment is running low on *rats,"* Omen calmly says out loud. Putting the gun to his cheek as if it were a pen, and he was thinking of a long math equation.

"Honor died when the meteors hit. It crushed every last ounce of the little honor left in this world. My favorite rat of chaos is dead, too. With the Beast dead, the sentinels have no leader…"

There was a slight pause as he sighed. Resting the gun on his shoulder. "You did come back. And I don't want to clean up the floor of your matter. Your survival lies in this last chance," he sighs.

"Thank you, sir," Silvia bowed.

"You will be under Angel's command."

"Sir… if I may remind you, Angel is thirteen."

"You have your orders." He sternly says, throwing the gun back on the desk. "If i may ask, the reports show that Beast was killed in hand-to-hand combat?"

"Reports show that a soldier infiltrated the mammoth and killed everyone inside, leaving Trevor for last and facing him in hand-to-hand combat."

"One rat destroyed my masterpiece of a machine?" Omen asks in a tone only described as bewildered.

"Yes, sir. His name is Ironsights. Nicknames vary from "Irons" all the way as chaotic as "the God of revenge." He is a soldier who fights for the Red Army. Confirmed kills reach the hundreds, and recent reports show that he is a commander now."

"A rebellious rat," Omen says with interest, talking to himself. "Of course, it is a shame he is not with us."

000